Microcomputer-Aided Maintenance Management

INDUSTRIAL ENGINEERING

A Series of Reference Books and Textbooks

Editor

WILBUR MEIER, JR.
Dean, College of Engineering
The Pennsylvania State University
University Park, Pennsylvania

Volume 1: Optimization Algorithms for Networks and Graphs, *Edward Minieka*

Volume 2: Operations Research Support Methodology, *edited by Albert G. Holzman*

Volume 3: MOST Work Measurement Systems, *Kjell B. Zandin*

Volume 4: Optimization of Systems Reliability, *Frank A. Tillman, Ching-Lai Hwang, and Way Kuo*

Volume 5: Managing Work-In-Process Inventory, *Kenneth Kivenko*

Volume 6: Mathematical Programming for Operations Researchers and Computer Scientists, *edited by Albert G. Holzman*

Volume 7: Practical Quality Management in the Chemical Process Industry, *Morton E. Bader*

Volume 8: Quality Assurance in Research and Development, *George W. Roberts*

Volume 9: Computer-Aided Facilities Planning, *H. Lee Hales*

Volume 10: Quality Control, Reliability, and Engineering Design, *Balbir S. Dhillon*

Volume 11: Engineering Maintenance Management, *Benjamin W. Niebel*

Volume 12: Manufacturing Planning: Key to Improving Industrial Productivity, *Kelvin F. Cross*

Volume 13: Microcomputer-Aided Maintenance Management, *Kishan Bagadia*

Additional Volumes in Preparation

Microcomputer-Aided Maintenance Management

Kishan Bagadia
UNIK Associates
Brookfield, Wisconsin

Marcel Dekker, Inc. New York and Basel

Library of Congress Cataloging-in-Publication Data

Bagadia, Kishan
Microcomputer-aided maintenance management.

(Industrial engineering; v. 13)
Bibliography: p.
Includes index.
1. Plant maintenance--Data processing. 2. Industrial equipment--Maintenance and repair--Data processing. I. Title. II. Title: Micro-computer-aided maintenance management. III. Series.
TS192.B3 1987 658.2'0028'5 86-23984
ISBN 0-8247-7553-8

MARCEL DEKKER, INC.
270 Madison Avenue, New York, New York 10016

Current printing (last digit):
10 9 8 7 6 5 4 3 2 1

PRINTED IN THE UNITED STATES OF AMERICA

Preface

This book is intended to be a comprehensive practical reference on computerized maintenance management for plant engineering personnel and others concerned with maintenance management. It applies to industrial plants, hospitals, buildings and grounds, and so on. Because of its logical development of topics and the integration of both theory and practice, this book should also prove useful in maintenance engineering courses and seminars conducted for professional groups in industry.

Management has realized that there is a need for better planning and control of maintenance. Whereas in the past operations were managed manually, computerized techniques are now available that can deal with a broad range of maintenance functions, for example, maintenance scheduling, manpower planning and forecasting, and purchasing of spare parts. Storing enormous amounts of data and generating useful management reports are important parts of such techniques.

The book treats independently each of the various components of maintenance management, such as equipment data, preventive maintenance, work order systems, and spare parts inventory control. The basics of microcomputers are discussed to make a novice user familiar with hardware components. Software selection criteria are discussed at length to help plant managers make an intelligent decision while selecting a software package for their application. It is my experience, in talking to hundreds of plant engineering personnel, that they want a computerized maintenance system but have difficulties in justifying one. A chapter has been included on justifying the system. The final step is to implement the project, which is also discussed.

The contributions of many people and organizations are included in this book. Whenever possible, an attempt has been made to accurately credit the sources of materials. Appreciation must be expressed to all who have helped, especially my office staff, who translated my scribbles into a presentable manuscript. Last, but by no means least, my thanks are due my wife and family, who cheerfully put up with me, and without me, as the book took shape.

Kishan Bagadia

Contents

Microcomputer-Aided Maintenance Management

1

Introduction

WHAT IS MAINTENANCE MANAGEMENT?

First, let us define maintenance. Maintenance can be defined as the orderly control of activities required to keep a facility in as-built condition, continuing to have its original productive capacity.

Maintenance activities are a major cost factor in most industries, affecting both return on capital and production throughput. Maintenance activities are usually categorized on the basis of when the work must be done. For example, emergency maintenance means that the work must be done immediately and preventive maintenance refers to activities that are carried out according to a planned schedule.

Maintenance management can be simply defined as managing the maintenance activities. A good maintenance management system makes equipment and facilities available. What does this availability mean? It means that production can get any item, such as light, power, air, gas, heating, cooling, or machine tools, when it is needed. If the equipment or service asked for is down, or if the machine stops short of a job completion, time and money are wasted. It is obvious that downtime has to be kept to a minimum, and a good maintenance management system helps accomplish that.

The fundamentals of maintenance management are not complicated. The basic steps of a maintenance management program are request, approve, plan, schedule, perform work, record data, account for costs, develop management information, update equipment history, and provide management control reports (see Figure 1.1). A brief review of the events that occur during each of these steps follows:

Request: Requests to perform maintenance work may be transmitted in different ways: verbal, telephone call, written request.

Approve: Simple jobs (small expenditure) may be handled by the maintenance supervisor. When large expenditures are involved, several levels of management approvals may be required.

Plan: The planning step can take many forms:
The maintenance supervisor may just give verbal orders to craftsmen.
If a planner is used, he will prepare a work order. On major projects, the planner may follow the critical path method (CPM), detailing all the steps, procedures, and instructions for carrying out the plan.

Schedule: Scheduling involves three factors:
Priorities: Priority codes based on established criteria or on the importance of the equipment and the kind of work to be performed may be used.
Job assignment: May be on a first-come basis or may reflect skill requirements and time estimates as part of a manpower scheduling plan.

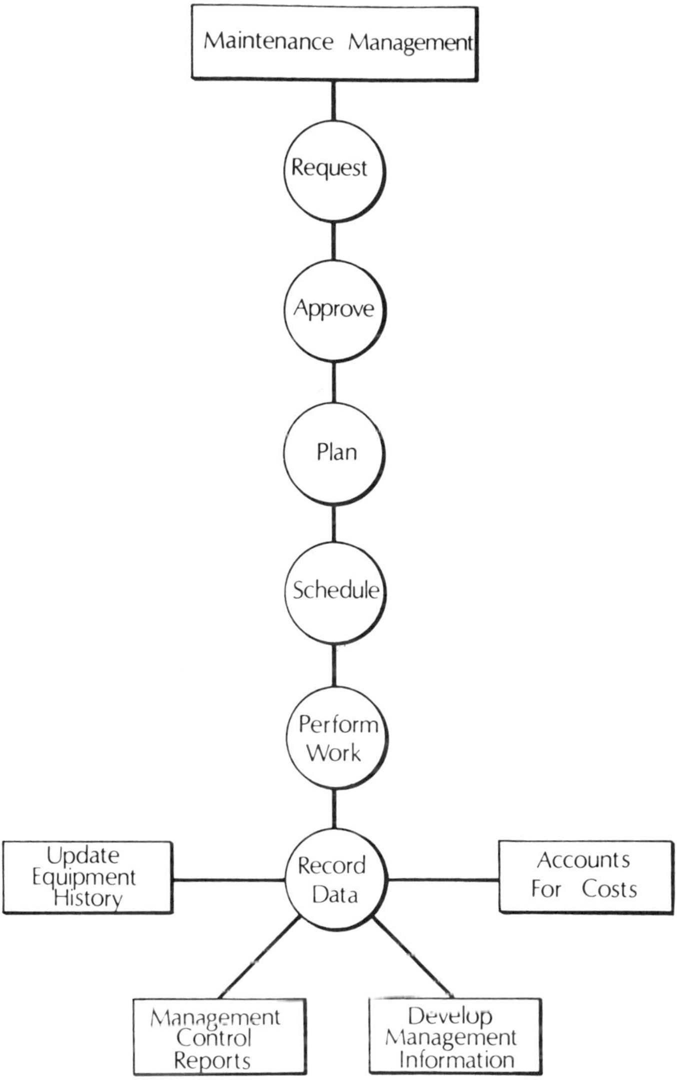

Figure 1.1 Maintenance management—basic steps.

Follow-up: May be a casual inquiry on when the work was completed or may involve feedback leading to a critical review of the job.

Perform work: This step involves craftsmen working from little or no instructions or from detailed procedures.

Record data: The data recording may vary from simply listing the actual hours to keeping comprehensive records of material charges, equipment identification, work assigned and performed, and other pertinent data.

Account for costs: It is important to know where and what the money is being spent for.

Develop management information: This may involve providing facts on current work, including costs, accumulated data, equipment identification, productivity, budgets, and scheduling.

Update equipment history: History records may vary from little or no data to on-line upgrading of all equipment, showing use, downtime, and maintenance labor and material costs on each piece of equipment.

Management control reports: As management information is developed, control reports covering expenditures, performance, backlog, equipment data, and so on, can be generated regularly to summarize the results of the maintenance function. These reports are very important for plant managers to make appropriate and intelligent decisions.

PROBLEMS

The key problems pertaining to maintenance management are as follows:

Manpower requirements: How many total craftsmen do you need in the maintenance department and how many in each maintenance category?

Performance
Employee productivity—How productive are maintenance craftsmen?
Machine productivity—How much is the downtime?

Planning to see that materials and tools are available to perform the job

Lack of history records

Incomplete material control records

Lack of preventive maintenance that causes high unscheduled downtime

Studies have shown that the productivity of most maintenance employees is only 30–50% (based on estimated and actual hours), often through no fault of the men. Obsolete maintenance management practices are the main causes and make it difficult for employees to do their job effectively. Delays that prevent craftsmen from achieving results are frustrating and costly. Employees spend a significant part of their time waiting or on nonproductive tasks, such as:

Looking for foremen to get job assignment

Visiting work site to find out what must be done

Rounding up materials and making multiple trips to the warehouse

Return trips for tools

Spare parts out of stock, no information on substitute parts

Waiting for other crafts to finish

Shortage of crafts to finish parts of the job

Waiting for a shutdown, clearances, and access to job site

Waiting due to lack of information or drawings

Losing time because of canceled orders

Rescheduled job priorities

Extended coffee or lunch breaks

Late startup, early washup

Waiting for special tools or engineering specifications

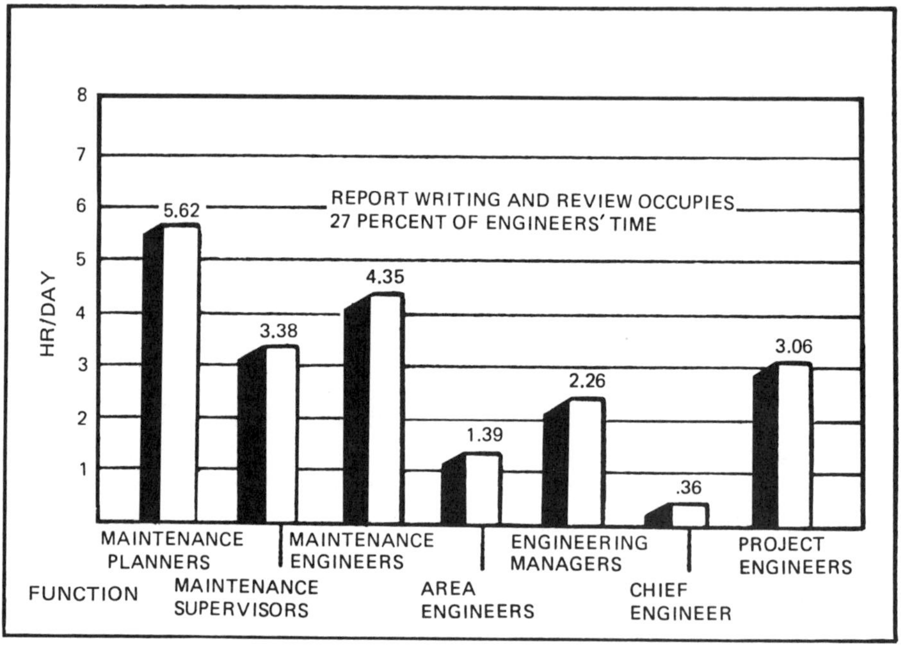

Figure 1.2 Time spent on paperwork in 500-person plant engineering organization (From J. Johnstone, *Plant Engineering*, Nov. 15, 1984, p. 83.)

Paperwork

The keystone of the maintenance organization is the foreman. Foremen are too important to spend time writing requisitions, looking for parts, and doing other nonproductive tasks. In many cases, the paperwork also requires a significant amount of wasted motion in carrying the information from one area to another. Paperwork also involves a significant amount of clerical time. Paperwork also requires plenty of storage space.

Figure 1.2 shows a typical example of how much time is spent on paperwork.

MANAGEMENT SUPPORT

Maintenance budgets and expense are typically considered necessary evils for the plant manager. Plant managers review the maintenance budget with an eye toward making it as small as possible each year. Some companies' maintenance departments operate smoothly and efficiently whereas others function with great difficulty. Why? Those that work have both management support and a reasonable budget. Every maintenance operation—small, medium, or large—needs these two components. It all boils down to a simple maintenance equation:

> Management support + reasonable budgets = Uptime = Equipment and facilities available

Maintenance must be treated as an equal in the whole manufacturing process: not as an excuse for lack of production, not as a necessary evil.

The management can support maintenance by (a) taking a positive attitude toward maintenance, (b) making maintenance part of the team, and (c) selecting maintenance personnel with equal care.

No department can succeed without management's support and a budget equal to the department's task.

In summary:

> The maintenance operation of the plant should be considered a profit contributor.
>
> Maintenance should be managed and measured as a profit contributor.
>
> Profits resulting from an improved maintenance strategy can be developed in a relatively short period by dramatically reducing downtime.
>
> The decision to address maintenance as a profit source is good for business, and it tends to build better team relationships between production operations and maintenance.

The question then arises, how do you get the support from top management? A number of steps can be implemented that will improve the credibility of the maintenance department in the eyes of top management. These include the following:

Show the maintenance department's capability and credibility by periodically reporting what it does and how effectively.

Speak and report in terms of dollars, since this is the language management understands.

Establish a baseline of present conditions and show how improvements are being made. If goals have been established, show how you are contributing and making progress toward them.

Provide management with realistic, well-thought-out proposals for: cutting production costs, increasing uptime, increasing productivity, etc.

Be sure proposals are well planned and professional in approach and style, and include technical feasibility and cost justification, as well as return on investment.

Take the initiative and keep the plant at a high level of availability. Planning, scheduling, and control are paramount. Avoid passive maintenance attitudes. Be aggressive in your approach to maintenance. Keep reaction maintenance to a minimum.

COMPUTERIZED MAINTENANCE MANAGEMENT SYSTEM

Probably few places, if any in the world today, have not been touched by high technology in some way. High technology is a phenomenon that is bound to bring about profound changes in the way all of us live, work, and play. For plant engineering, it could mean an era of increased responsibility and status, but only if plant engineers recognize the opportunities offered and seize the challenges presented. If American industry is to regain its competitive vitality, it's going to have to take a massive injection of high technology.

Maintenance management software packages are somewhat familiar to most plant engineers. They can be used to plan, schedule, and control, and to provide real-time data on maintenance activities to achieve a predetermined maintenance level. They can also be used for inventory analysis and control.

The importance of maintenance is receiving greater recognition as more manufacturers adopt material requirements planning (MRP or MRPII) systems.

It is important that scheduled maintenance be a part of these programs. In theory, the master schedule drives purchasing, which in turn drives production activity. In reality, on a breakdown maintenance program, it is the breakdowns that drive a repair/parts purchasing schedule, which in turn pushes production activity. Scheduling outages is more important to MRP, and software programs that can control maintenance activity are the answer to accurately meeting those schedules.

The see-feel-and-touch aspects of maintenance are slowly giving way to electronic information gathering as the need for that information becomes more demanding. Systems on the market can call up vast amounts of information about a particular machine in just a few seconds. The maintenance supervisor might need a complete maintenance history on a particular machine, or he might need to know the status of machines waiting for repair parts. All of this information is immediately available. This kind of information gathering has been available in other areas of manufacturing and is now becoming part of maintenance programs as companies work toward a total electronic operation.

As automation drives up the cost of downtime, comprehensive predictive maintenance packages will become more economical. These packages allow the computer to monitor plant systems, checking status of bearings, vibration, power consumption, and so on, and then to schedule and order maintenance at the most economical time. Such a system issues work orders, prints out detailed instructions for workers, provides all necessary drawings, orders tools and parts from stores, checks results, reports costs, and more.

Until recently, most maintenance management information systems were programmed for a mainframe computer. As a result of this, maintenance departments had to share the equipment, programmers, and overhead with others in the plant. With the recent drop in the cost of microcomputers, very small companies can now afford one for maintenance application. Even plants with existing mainframe computers can afford to purchase a micro and devote its full-time use to the maintenance management information system.

Manual systems of maintenance cannot compare with the accuracy, efficiency, speed, and sophistication of computer technology. The recent rise in the number of computerized maintenance software packages is comparable to the growth of word-processing and accounting software in the last few years. And, like word processing and accounting, computerizing maintenance tasks allows the paperwork, which historically was a manual function, to become an automatic one.

Too many companies have a negative attitude toward the cost of maintenance. They view it as an inevitable expense about which nothing can be done. And so they continue to employ outdated and inefficient methods of maintenance management. Maintenance is an inevitable expense, but it can be reduced dramatically. Through a system of computerized maintenance management, the useful life of equipment can be prolonged and cash crunch caused by replacement can be alleviated. When there are fewer breakdowns, there are fewer work interruptions, and as a result, maintenance productivity, along with plant productivity, increases.

Maintenance costs can significantly affect a company's profit-and-loss statement. That is why the plant engineer and his maintenance manager must have accurate, timely information relating to labor, materials, and inventory costs involved in keeping the plant operating. In many cases, this cost information can only be managed economically with the aid of a computer.

Each maintenance cost should be available to the maintenance manager so that he can tell where he is in relation to the budget. And each maintenance cost should be broken down as it relates to costs for work order processing, emergency repairs, preventive maintenance, labor and materials, and total inventory. If a plant engineer or maintenance manager is not in a position to obtain verifiable figures when needed, he does not have good control over the maintenance operations.

Maintenance departments are faced with other problems besides costs. In the past, most maintenance departments could keep one step ahead of possible disasters with informal management of paperwork, schedules, and follow-up. Today, maintenance is feeling the weight of more and more paperwork; at the same time, it is expected to maintain more complex production machines, keep a larger inventory of spare parts, and do these tasks efficiently. If the plant engineer is unable to keep ahead of all the demands placed on his operation, he should consider using computers. They can provide the means to keep closer control over all functions.

The following sections describe objectives, features, and benefits of a computerized maintenance management system.

OBJECTIVES

1. Every part of the program must have the full acceptance of the maintenance department, which would control and operate the system.
2. The program should be readily adaptable so existing plant facilities and equipment as well as current maintenance procedures.
3. The program must have full support of the upper management.
4. It must contain all the elements of a good
   a. preventive maintenance program
   b. work order system

   c. equipment history system
   d. inventory control system
5. The program information must be readily accessible to all maintenance supervisors and management.
6. It must be designed to allow continual updating.
7. It must be simple and understandable to all levels in the organization. The system should be designed for the end users, not for the computer experts.
8. a. Clerical functions and paperwork must be minimized without sacrificing effectiveness.
   b. The foreman's paperwork must be reduced so he can spend more time supervising work and solving specific plant problems.
9. Maintenance data and information must be as complete as possible.
10. The opportunity to start any new facilities with this maintenance program should be retained.

FEATURES

1. The maintenance management system should be totally integrated, designed to be used by all levels of personnel at the plant (see Figure 1.3). The integrated system means information entered at one point would be automatically and instantly available at all other appropriate places throughout the system.
2. The key sections to the program should include:

   Equipment identification

   Referencing and cross-referencing

   Complete and comprehensive equipment specifications

   Spare-parts list for each piece of equipment

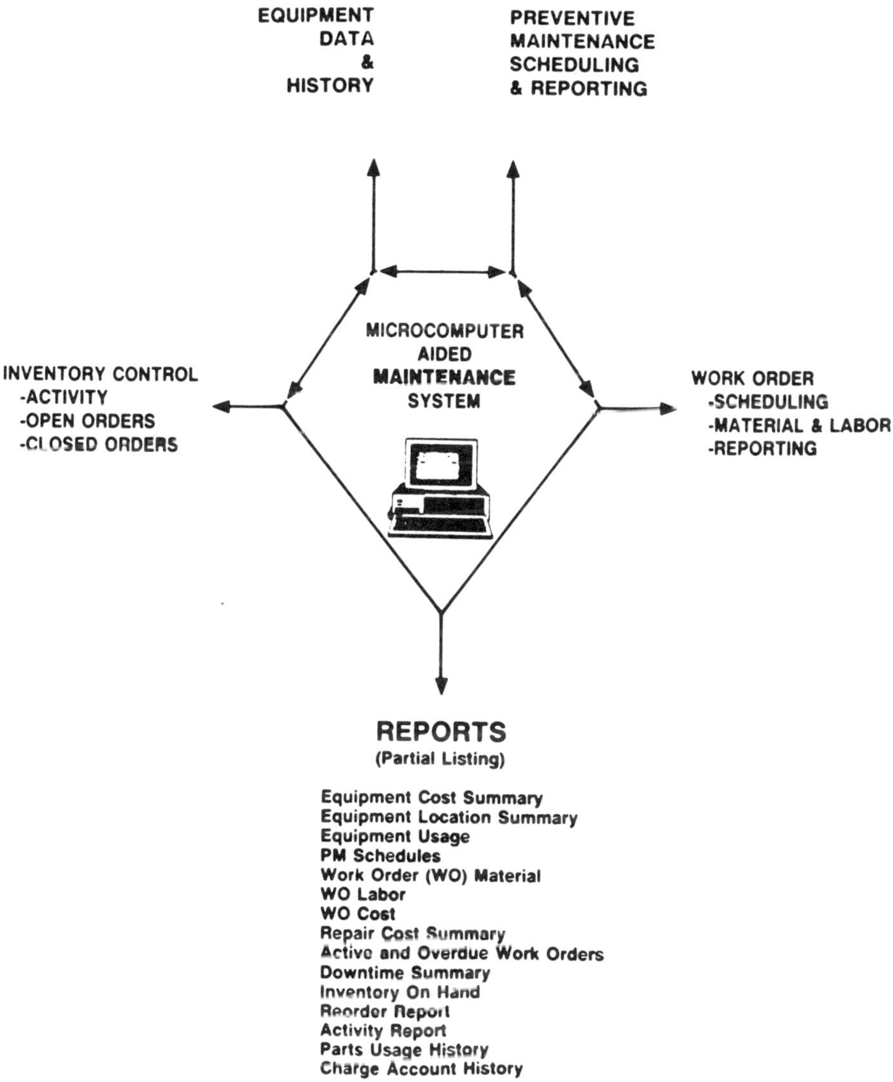

Figure 1.3 Microcomputer-aided maintenance management system.

Warehouse inventory levels and purchasing

Preventive maintenance worksheets and checklists

Preventive maintenance schedules

Work order preparation

Outstanding work order listings and priorities

Maintenance histories and downtime records

Manpower summaries

Equipment listing

Work order inquiry

Equipment cost histories

Major overhaul procedures and standards

Uniform method to initiate work

Procedure to approve and authorize work

Weekly review of work priorities

Weekly work program based on craft net capacity to do work

Uniform methods to measure and control work backlog

Recording man-hours, materials, and job costs

Uniform procedure to report completed work

3. The system should provide the maintenance personnel with:
   a. a clear picture of what is expected of them
   b. adequate feedback on their performance
4. Interface with existing plant data-processing programs where possible and necessary.

BENEFITS

1. Increased productivity of the maintenance work force.
2. Reduction of maintenance costs.

3. Reduction of overtime.
4. Reduction of outside contract work.
5. Reduced maintenance backlog.
6. Reduction of downtime.
7. Reduced cost per repair.
8. Employee frustration will be reduced resulting in improved morale. A happy worker is a productive worker.
9. Better service to other departments.
10. Reduced inventory costs.
11. Paperwork will be significantly reduced to help make possible the most productive use of employee time.
12. Access to maintenance information is dramatically expanded.
13. Maintenance and supervisory personnel can use their time more efficiently.
14. The computer system is effective in showing individuals and crews what they are contributing to the operation. Each craftsperson knows what impact his or her efforts have on the operation.
15. Longer useful life for production equipment.
16. Decreased delay in systematic planning, scheduling, and control by making material and special tools available when needed, providing clear instructions, providing up-to-date history records.
17. Use of historical maintenance information to make better capital expenditure decisions in relation to older equipment, that is, to help decide on repair versus replacement of specific equipment.
18. Reduced follow-up role required of the skilled trades supervisor.
19. Store inventories can be made more effective by expanding recorded information to include such items as

vendor lead time, delivery, reorder points, maximum inventory levels, and economical order quantities. As a result of these, the right parts or materials are more likely to be in the right place at the right time.

20. *Increase in overall plant productivity.* It is hard to cite the exact percentage of productivity improvement as each organization is different. However, if the system is planned appropriately, significant improvement could be realized.

21. *Save money (more profit for the company).*

2

Understanding Microcomputers

INTRODUCTION

This chapter is written to aid plant engineering personnel in understanding microcomputers. We are entering a new era—the information age—where the computer will have a great impact on our society. In the past, significant changes in the way business was conducted occurred over decades. It is predicted, however, that with the microcomputer, significant changes will result in less than one decade. The objective of this chapter is not to write a comprehensive text on computers; rather it is to prepare maintenance managers to make an informed and intelligent decision about purchasing a microcomputer.

In 1947 a fully electronic computer had 18,000 tubes, weighed 30 tons, and cost $500,000. Today a microcomputer with

20 times the computational power has been reduced in size to a ¼-in. square silicon chip costing $10.00. Technological advances such as this continue, and the price of computer systems continues to decline—to the point that, at some time in the near future, a computer will be as much a part of running a business as are a telephone and a cash register.

The development of the microcomputer occurred recently, in the early 1970s, and related innovations are occurring at a rapid rate. Unless an individual works in an environment directly related to this new technology, it is nearly impossible to keep up with new developments.

This is particularly true of the maintenance personnel who spend long hours managing their department. They cannot, however, afford to ignore this new technology because it can have a large impact on how they manage their business.

TYPES OF COMPUTERS

Basically, there are three types of computers: (a) mainframe computers, (b) minicomputers, and (c) microcomputers. Mainframe computers are for use by large numbers of people at the same time, approximately 20–100 users. The cost ranges between $250,000 and $1,000,000. Minicomputers serve from 4 to 40 users. The cost ranges from $50,000 to $250,000. Microcomputers serve from 1 to 10 users and cost range is $2000 to $25,000.

CLASSIFICATION OF MICROCOMPUTERS

Microcomputers can be classified by size, configuration, function, computing power, and architecture. There is no standard approach.

Some of the more common microcomputer types are the following: (a) handheld, (b) portable, (c) transportable, (d) desktop, (e) rack-mounted, (f) personal, (g) multiuser, (h) supermicros, and (i) business-oriented.

We will concentrate on personal and multiuser microcomputers because they are the best suited for maintenance applications.

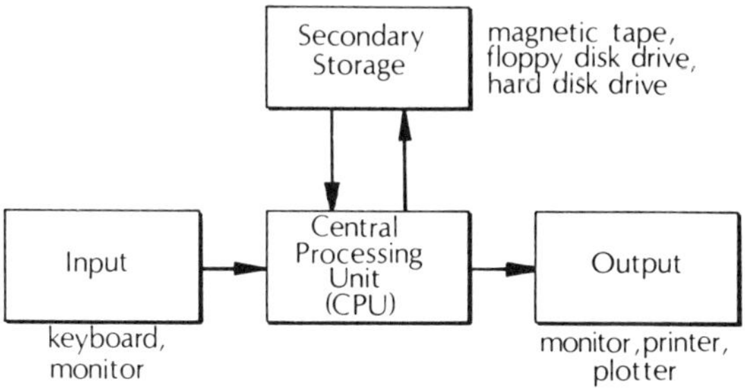

Figure 2.1 Major computer components.

Personal Computers

The personal computers are microcomputers designed to be used by one person. The term *personal computer* refers to a general-purpose, desktop business machine capable of running popular business software and engineering calculation packages.

Multiuser Microcomputers

The multiuser microcomputer has multiple terminals attached to a single electronics unit that contains the same type of microprocessor used in standard personal computers.

Components of a Microcomputer

A computer receives, stores, manipulates, and communicates information. It has four major components: central processing unit (CPU), storage facility, and at least one input and one output (I/O) device (see Figure 2.1).

COMMONLY USED TERMS

Central processing unit (CPU): The heart of the system unit is the "mother" board, a printed circuit board containing the microprocessor, CPU chip, associated electronics, and usually a bank of memory chips. This board has several slots that accept plug-in "daughter" boards for controlling peripheral hardware such as a printer.

Memory: Any device into which a unit of information can be copied, which will hold this information, and from which the information can be obtained at a later time.

RAM: Random-access memory (RAM) is volatile memory, lost each time the computer is turned off. The usefulness of a computer is a function of the amount of RAM available. Sufficient space must be available to store the operating system software and the application program and leave room to store data for processing. RAM is usually supplied by 16- or 64-byte chips. Most microcomputers come with 64K bytes RAM as standard, but this figure can be expanded to 256K or 512K bytes in most 16-bit computers by installing more memory chips.

ROM: Read-only memory (ROM) is nonvolatile memory. Typically, ROM is used only for loading the operating system from disk into RAM.

Alphabetic character: A character that belongs to the set of letters, A B C D E F G H I J K L M N O P Q R S T U V W X Y Z, and the space character.

Alphanumeric: Pertains to a character set that contains both letters and numerals, and usually other special characters.

ANSI (The American National Standards Institute): Composed of representatives from industrial firms, technical societies, consumer organizations, and government agencies. This group develops and approves such things as technical terminology, symbols, abbreviations, coding structures, performance characteristics, methods of rating, etc. Because the organization was known formerly as the United States of America Standards Institute, this group is sometimes referred to as USASI.

ASCII (American Standards Code for Information Interchange): A uniform code in which alphabetic, numeric, and special characters plus several special symbols are represented by eight-bit configurations.

Backup: A copy of a disk or tape to allow recovery of data in case of operator error or equipment failure.

Bit (Binary Digit): The smallest element of binary machine language represented by a magnetized spot of a recording surface or a switch being off or on. The state of off or on or magnetization is represented by 0's or 1's.

Byte: A sequence of eight adjacent bits operated on as a unit, may be shorter than a word.

CRT (cathode ray tube): A television-type device used as an output terminal for computer systems.

Daisy wheel: A printer type; characters are on the ends of spokes of a wheel, like petals on a daisy.

Data: A collection of facts or figures.

Database: An organized set of data stored in the computer's secondary storage (tape or disk).

File: A collection of records, an organized collection of information directed toward some purpose. Normally resides on a secondary storage device, such as a disk or a tape.

Flow chart: A graphical representation of a sequence of operations using symbols to represent the operations.

Hard copy: A printed copy of machine output in a readable form, such as output from a printer.

Input: The data to be processed.

I/O: Input/output.

Modem: A device that provides the appropriate interface between a data-processing machine and a communications line. It converts data originating in digital form into analog signals suitable for transmission over telephone lines (and vice versa).

Output: The transfer or exit of processed or unprocessed information from a system, location, or area to a destination or external point.

Program: A sequenced set of instructions to a computer to do a particular job.

Terminal: A point in a system or communication network at which data can either enter or leave.

PERIPHERAL EQUIPMENT AND DEVICES

Input Devices

Keyboards are the basic computer input device. Some keyboard layouts are better than others, but the user must generally accept the one supplied with the computer because few companies offer options.

Light pens contain a light sensor that picks up the scanning beam of the CRT when the pen is activated by pressing it against the screen. A timing circuit translates the position of the pen. Pointing with a light pen is a natural hand movement, but the arm becomes tired holding the pen against the screen (a pen on a table-top digitizer pad is less tiring).

Mouse is the name given to the palm-sized device that, when moved across the table, positions the cursor on the CRT. There are two general types: mechanical and optical. Mechanical versions house a ball that drives two shafts when it is moved; the shafts actuate mechanical switches, potentiometers, or optical sensors to register movement. Optical versions, containing one or two light-emitting diodes and detectors, are moved across a mirrored pad containing a nonreflecting grid; the pattern of reflected signals registers movement.

Other input devices are *video digitizers* and *digitizer tablets.*

Mass Storage

Media for saving data and programs include magnetic tape, flexible (floppy) magnetic disks, and hard magnetic disks.

Magnetic tape is not a popular primary recording medium for microcomputers. However, it is often used to back up hard disks.

Floppy diskettes are the primary storage medium for microcomputers. They come in different sizes (3½-, 5¼- and 8-in. diameter) and densities (single and double sided, and single, double, and quadruple density). Most personal computers use the 5¼-in. size, but 3½-in. drives are starting to appear in some of the new models. The 8-in. size is normally used on larger business systems. Recording density is a function of the disk drive hardware, and format is a function of the operating system. A 5¼-in. DSDD (double-sided, double-density) floppy disk can store about 360K bytes of data. Diskettes recorded on one brand of computer cannot generally be read by another brand, because they use different formats (the recording pattern of tracks and sectors), but the diskettes themselves are interchangeable.

Winchester hard disks use the same type of recording medium as floppy diskettes but it is bonded to a metal substrate and usually sealed in the drive unit. They can store 5–140M bytes of information, but the 10M-byte size is the most popular. A 5¼-in. Winchester disk can be mounted in the same space required by a 5¼-in. floppy disk drive. Hard disks store more information than floppy diskettes, and information can be recovered more quickly because they spin at higher speeds. Hard disks are especially useful when users need to switch frequently from one program to another and for operations that take massive amounts of data, such as maintenance histories of plant equipment.

There are other storage devices also, such as *disk emulators* *bubble memory*, and *EPROM* (erasable programmable read-only memory).

Monitors

Monochrome monitors, primarily used for text, should be capable of displaying 25 80-character lines (some can display 132-character lines).

Color monitors: *Composite video type* color monitors generally do not have sufficient resolution for business graphics and text display. *RGB monitors*, with high resolutions, work well with most applications. *Touch-screen monitors* have touch-sensitive screens that allow the user to move the cursor by touching the screen with a finger.

Printers and Plotters

Dot matrix impact printers can print characters and graphics patterns. The image is formed by the impact of individual wires on an inked ribbon. Software, part of which resides in the printer, determines the pattern of dots. Different type styles (enlarged, condensed, boldface, italic) can be produced by software rather than different print heads. Operating speeds vary from 30 to 200 characters per second (cps), depending on mechanical design and style of type being printed. Color printing is possible using a three-color ribbon, but the price is high. Dot matrix printers are noisy.

Dot matrix nonimpact printers use heat or electric discharge from pins on coated paper rather than ink on plain paper to form the dot image. Nonimpact printers are quiet, but they cannot make carbon copies, and the coated paper may be annoying. Printing speed for thermal models is about 40 cps.

Ink jet printers are a form of nonimpact dot matrix printer in which matrix dots are formed by spraying a drop or a jet of ink on plain paper. They are fast and quiet, but expensive. Printing speed is around 150 cps.

Letter-quality printer is a general term applied to an impact printer with fully formed characters. It works like a typewriter; characters are cast or embossed on a daisy wheel or thimble-shaped printing element (ball-type printing elements are too slow). Type styles are changed by changing the printing element. They are noisy. Printing speed varies from 12 to 80 cps.

Plotters use colored pens to transfer graphics data to paper or film. The plotter holds one or several pens. One type moves the paper or film under the pen on one axis and moves the pen across the paper along the other axis to complete the desired plot. Only the pen is moved on other types. Software selects the pens and controls the lifting and placement of them on the paper. Plotters are available for all standard drawing sizes.

SOFTWARE FOR MICROCOMPUTERS

A computer is worthless without software to make it work. It cannot even be used for writing general programs, unless it has a special boot or primer program (supplied by the manufacturer in ROM) to get it started. Once underway, it still cannot do much without a housekeeping and coordinating program called an operating system. It also needs an application program with the instructions needed to accomplish useful work and a translator program for converting user written programs into code understood by the computer.

SELECTING A SYSTEM

Putting together a microcomputer system is different from other types of equipment selection projects because the field is changing so rapidly, and most personal computers are sold through retail computer stores. Store personnel are typically strong on computer knowledge but weak on engineering application know-how. Maintenance managers must do their homework first, defining their needs in great detail before shopping for hardware and software.

The rule has been to select software first and then pick a computer that can run it. This rule is still valid, but less important now that most software designers have multiple versions of their products, one for each major type of equipment.

Whether purchasing hardware or software, one should determine the level of support available from both the selling organization and the manufacturer.

Computer maintenance can be a problem. Computers are not normally repaired by plant personnel, and many stores offer minimal service. Because sending a computer back to the manufacturer for repair is often not practical, one should consider a service agreement by which the computer manufacturer, store, or user contracts with a third party for maintenance and repair.

Some other factors to consider in selection of hardware include the following:

Will it do the job you want it to do?

Does it have the necessary power and capacity?

Does it allow for future expansion?

How long has the vendor been in business?

Does it include a backup device.

When was the first hardware system installed?

What number of hardware systems have been installed?

The software selection procedure is covered in detail in Chapter 6.

CONCLUSION

You now have some basic information about microcomputers—the hardware, components, and selection. This will help you go to the dealer and software vendors, sort out their claims, and make a decision about what is the best system for you.

3

Preventive Maintenance and Equipment Data

INTRODUCTION

Preventive maintenance (PM) means different things to different people. Management frequently sees it as "overhead that may or may not be of value"; production people usually consider it to be "something that prohibits scheduled product delivery"; line mechanics place it somewhere between corporal punishment and management stupidity. In general, their attitudes are summed up by: "The machine is working fine, why work on it?"

The answer many plant engineers give to the question "Do you have a preventive maintenance program? Is "Yes, we have a PM program." However, each plant engineer may have a completely different idea of just what a PM program is, the important ingredients of a good system, and the results that can be expected from a well-planned program.

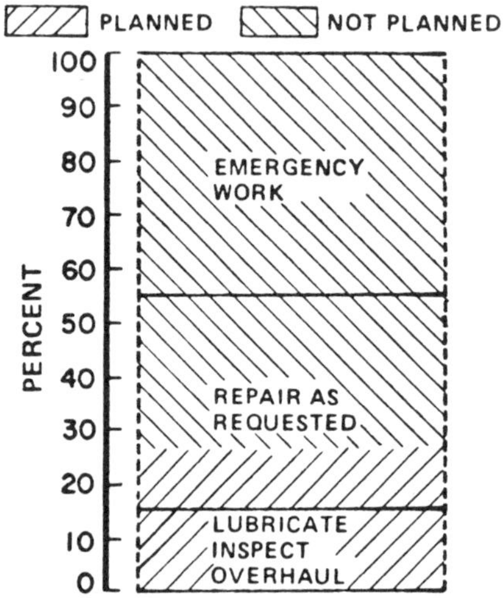

Figure 3.1 Distribution of maintenance hours. (From E. T. Newbrough, *Plant Engineering*, March 6, 1980, p. 291.)

Typical comments explaining their PM programs can include:

"We have two full-time lube men."

"We inspect our equipment on regular schedules."

"Our mobile equipment is serviced routinely."

"Our mills are overhauled every 6000 operating hours."

"We schedule downtime for preventive maintenance every week."

"We do all PM during the 2-week annual shutdown period."

The answer to the question of what percentage of the maintenance hours is spent on preventive maintenance is usually 5 or 10%—seldom more than 15%. If this estimate is accurate, 85% of maintenance hours in industrial plants may be unplanned. And unplanned maintenance is expensive; it costs industry hundreds of millions of dollars each year. (See Figure 3.1.)

In the majority of plants today, we find two types of maintenance. Both start with the letter "P," but there the similarity ends. One is "PANIC" maintenance, which is mass confusion with no control. The other is "PREVENTIVE" maintenance, which is organized with a well-trained staff and controlled. The latter type is what we will cover in this chapter.

Management must review the overall facility operating procedure and conditions, analyze company goals, and then decide whether to: (a) continually operate all facilities and equipment at maximum capability, shutting down only for emergency repairs, or (b) employ a PM program on a firm, regular schedule, perhaps reducing daily plant output slightly but, in turn, practically eliminating costly plant shutdowns.

Preventive maintenance should be considered in the same light as an insurance policy. Companies regularly buy thousands and even millions of dollars of protection against fire, explosion, wind, and other natural disasters. The major difference between PM and insurance is that insurance only pays off after the damage is done. Preventive maintenance protects a company against unexpected equipment breakdown that disrupts the production schedule and causes loss of business because of inability to deliver on time.

IMPORTANCE OF PREVENTIVE MAINTENANCE

The trend toward modern practice of organizing maintenance work to prevent interruptions of operations is the outgrowth of several factors:

1. *Increased mechanization*: Decreasing direct labor cost per unit of production by automation has required increased maintenance cost of the equipment. It is uneconomical to maintain large emergency maintenance staff which planning and systematic inspection can avoid.

2. *Close management of production*: Minimum stock between operations. Direct material flow has made interruptions more costly. Lost profits from production stoppage frequently exceed the cost of idle labor.
3. *Failure to deliver on time*: Serious consequences and possible loss of business may result from interruptions to operations.
4. *Correction of defective conditions*: Need to decrease cost of repairs and maintain performance efficiency of equipment and buildings.
5. *Planning of maintenance operations*. A well-planned preventive maintenance program will ensure availability of spare parts and reduce inventory cost (obsolescent material-insurance items).

OBJECTIVES

1. *Increase reliability of machines*: The primary objective of a PM program is to ensure that the machinery is functional and in good working order, or to ensure maximum productivity.
2. *Reduce downtime to a minimum*: Downtime has a direct effect on production. Productivity and profit decrease in direct proportion to the increase in downtime.
3. *Decrease maintenance costs*: With the installation of a good PM program costs can be reduced.
4. *Increase productivity (profit)*: If the first three objectives are met, it is a foregone conclusion that there will be an increase in production, with a corresponding increase in profits. Many articles are written about how we are not keeping up with foreign competition. One reason is that the competition has, and uses, preventive maintenance as a workable tool.

BENEFITS

1. Reduces number of emergency breakdowns.
2. Keeps optimum ratio of emergency versus planned work.
3. Provides information for considering replacement.
4. Improves customer service (better delivery, for example).
5. Reduces inventory cost by ensuring availability of spare parts.
6. Decreases maintenance costs. Once the program is established, an increase in PM planning effort often causes a decrease in maintenance costs. (The reduction in maintenance costs though, is not the primary reason for planning PM.) The savings in maintenance hours and overtime, although substantial, are dwarfed by the savings realized from a reduction in number of equipment failures. For many plants, having a process line down for only a few hours can cost more in lost revenues than the entire annual maintenance budget.
7. Increases overall productivity.

FEATURES

A good computerized PM system should include, but not be limited to, the following features:

Schedules work orders by calendar or metered usage or combination of both.

Permits scheduling of multiple work orders for equipment.

Prints PM work orders.

Provides a detailed description of PMs to be performed which are printed on work order.

Schedules multiple crafts for any PM.

Schedules the next PM based on the date the prior PM was completed.

Has the ability to schedule PMs for a specific date or day of the week.

Prints PM work load forecast for any specified period.

Has the ability to schedule PM for equipment and/or facilities.

PM schedule can be based on craft code.

Permits user to specify the lead time required to be notified of PMs to be generated.

PM Work Orders

Produces both PM and corrective work orders.

Produces a periodic exception report of past-due PM work orders.

Tracks labor and materials cost.

Produces a report of selected deferred work orders.

Has the ability to obtain work order backlog by craft code.

Upon request produces a report of active PM work orders.

Has the ability to select work orders by any status code or a combination of codes.

Maintains an on-line, detailed work order history by equipment for at least a year.

Automatically flags corrective work orders that are for equipment under warranty.

Each PM work order to be assigned a status code, which may be updated during the life of the active work order.

Reporting

The system should produce daily control reports which are printed on an exception basis.

Has a data inquiry capability to allow information to be selected based on data fields.

Prints reports that can be presented and sorted by data fields.

Prints equipment cost histories for any time period.

Automatically generates a management report at a specified interval (i.e., monthly).

Permits user-defined reports which can be reproduced by report name.

Produces cost reports to show the expense of using outside contractors and renting equipment.

Tracks equipment downtime.

Provides a work load plot by craft code to aid in resource leveling.

HOW TO START A COMPUTERIZED PM PROGRAM

A computer can be a very helpful tool in the management of a PM program, provided the needs of the maintenance function under consideration warrants the use of one.

Compiling the Data

The first step is to compile the data for each piece of equipment that is planned to be part of the computer system. Figure 3.2 shows a typical equipment data record, and Figure 3.3 shows PM requirements for a particular piece of equipment.

Only 3-4 hr per equipment item should be invested to compile all necessary data, read the maintenance operator's manual, and develop and write the programs. If similar types of equipment are involved, the job can be done faster. In such cases only the equipment data need be compiled; the PM requirements remain the same.

In compiling data, make maximum use of free information or services from equipment vendors and lubrication suppliers.

```
                              **** EQUIPMENT DATA ****
EQUIPMENT NUMBER: ===============                    MFR: ====================
            NAME: =====================       MODEL/TYPE: ===============
      DEPARTMENT: ==========                          SN: ===============
        LOCATION: ===============            MFR CONTACT: ====================
   SIZE/CAPACITY: ==========                       PHONE: ============
          RATING: ==========              VENDOR CONTACT: ====================
          WEIGHT: ==========                       PHONE: ============
            COST: ======0.00
  DATE INSTALLED: 00/00/00     CURRENT USAGE READING: ========0  UNITS: ======
 ═══DESCRIPTION═══════════════════════════════════════════════════════════════


 ═════════════════════════════════════════════════════════════════════════════
                           MOTOR MFR: ====================  VOLTS: =====
** ELECTRICAL DATA **     MODEL/TYPE: ===============        AMPS: =====
                                  SN: ===============         RPM: ===== HP: ====
```

Figure 3.2 Typical form for equipment data. (Courtesy of UNIK Associates; parts 1 and 2 have been filled out.)

When custom equipment has been purchased, the vendor is the logical source of information on the PM required. Vendors are usually willing to supply blueprints, as well as operating and maintenance manuals. In older plants, it may be necessary to write to vendors and pay a nominal sum for such manuals.

Strive for maximum involvement of company personnel. An obvious benefit to this concept is to minimize spending on outside services. Whatever can be accomplished internally without disrupting the overall operation benefits the company and permits plant personnel to claim ownership of the system. If they feel their input is needed, the chances for success are enhanced. A PM program should be viewed as a dynamic, not static, entity. As new techniques are learned, they should be added to the system. Many PM programs fail unless they are reviewed periodically and changes are made to update procedures or eliminate unnecessary steps.

Use clerical help for information gathering and record keeping. It is necessary to transfer information from manuals, specifications, and blueprints to equipment data and PM work order files.

```
                     **** PREVENTIVE MAINTENANCE ****

EQUIPMENT NUMBER: ===============          EQUIPMENT NAME:
    JOB NUMBER: ====    PRIORITY: ==            LOCATION:
 ACCOUNT NUMBER: ==========                   DEPARTMENT:
         CRAFT: ====     EST HRS: ===0.00   ACTION CODE: ====

     FREQUENCY: ===0 (DAYS)       ========0 (USAGE)        (UNITS)
      NEXT DUE: 00/00/00          ========0 (USAGE)

   PRINTING WO: Y (Y/N)    NUMBER OF DAYS TO ADVANCE WORK ORDER: =0

         WORK ORDER ACTIVE FOR THIS JOB:  * NONE *

==PM DESCRIPTION====================================================
```

Figure 3.3 Typical form for preventive maintenance. (Courtesy of UNIK Associates, parts 1 and 2 have been filled out.)

Consolidate practical experience data from mechanics, foremen, and so on. The person who works on the machine every day knows it. Besides, every maintenance department is always solving problems and developing new techniques to make equipment work more efficiently. It would be unfortunate not to identify and use those techniques in the formal PM program.

Developing a Priority List

Identify crucial equipment where (a) failure of the unit would endanger health or safety of personnel, (b) failure would stop significant part of production, and (c) the repair cost of breakdown is high.

If you have two pieces of equipment, one automatic, the other manual, obviously the automatic should come first. However, setting priorities is a problem that should be discussed with the production people since this concerns their operations.

Preparing a PM work program

Develop check sheets and inspection guides.

Develop PM frequencies for each task.

Assign crafts.

Develop a master schedule.

Scheduling Preventive Maintenance

Before PM can be scheduled by the computer, the various steps outlined previously must be taken. In addition, the following must also be done: (a) Ensure that work orders are written and approved for all PM work to be accomplished. (b) Ensure that manpower and materials required to form each element of PM work have been determined and provided for.

Once all of these steps have been taken, the necessary data can be entered into the computer. The computer can then be programmed to print out on a regular basis, usually once a week, the PM schedules previously determined. The once-a-week printout would include daily schedules and a weekly schedule, as well as the necessary work orders and instructions to go with the schedules. The computer can also be programmed to search the material inventory files and print out any necessary material requisitions, reserving the material as well. If in its search it determines material is not available, a separate printout can be made showing what is required. The work order requiring that material will also be flagged separately and not scheduled.

Preparing Reports

There are many reports that can be generated for use in controlling PM, but it is advisable to limit the number of reports to those which are most beneficial and will be used. The reports that are quite helpful in managing PM are the monthly cost center equipment maintenance report and the equipment maintenance cost record. (See Figures 4.4 and 4.5 in Chapter 4, "Work Order System.") The monthly cost center equipment maintenance report shows, by cost center, monthly maintenance costs of individual pieces of equipment, with year-to-date totals and the total for all of the previous year. By analyzing this report, one can spot a piece of equipment that appears to have excessive maintenance costs. Once a piece of equipment is so spotted, the computer, on request, will print out the equipment maintenance cost record for that piece of equipment along with a breakdown of all individual work orders for it. Some other reports that can be generated are: (a) the maintenance cost summary by cost center, which shows monthly maintenance costs broken down by class of work, including PM, with year-to-date totals for the cost center, (b) active PM work order report, which shows all PM work orders that are under process; and (c) overdue PM work order report, which shows all PM work orders that are past due.

4

Work Order System

INTRODUCTION

The maintenance manager or plant engineer begins to obtain the control necessary to genuinely manage maintenance when a work order system is put into effect. Without a well-structured, carefully adhered-to system, the maintenance organization is likely to find that the work coming in controls the organization rather than the other way around.

However, with a well-established work order system and management commitment to its use, a maintenance department will realize rewards of higher productivity, reduced costs, and increased pride in the maintenance function.

A work order system should produce quantifiable feedback on overall maintenance performance, job and plant costs, and equipment history. Maintenance should be able to give its customers accurate cost information on anticipated maintenance needs, so those using maintenance services can plan for maintenance expense.

The positive results achieved from a manual work order system are further enhanced by computerization, provided the computer system is chosen to meet clearly defined objectives and adequate planning is done for its installation and use. Computer use will be most rewarding for a maintenance department if they purchase equipment that is dedicated to maintenance use and not dependent on a mainframe system controlled by data processing or accounting.

Computerization should increase the efficiency of the maintenance operation and affect manpower, material, and equipment costs. The largest cost savings usually come from improved craft productivity, increased production uptime, and a reduction in inventory. Planned maintenance increases equipment life and minimizes unexpected breakdowns. Information stored by the computer allows faster and more accurate estimating of labor and materials needed to perform a job. Better management of resources results in a reduction in idle time and tool costs. Improved inventory control assures adequate quantities of needed spare parts and other materials, while unneeded items can be eliminated.

Quick access to accurate information speeds the maintenance organization's response to customer needs. With an on-line computer system, work orders can be written, planned, and scheduled at a CRT terminal. The information becomes immediately available on other terminals used by supervisors, craftsmen, or the customer. The computer can produce a printout of a daily job plan and the individual work orders making up the plan. All work orders in the system can be reviewed and updated at any time. Backlogs and craft availability can be monitored easily.

Successful use of any computer system requires a commitment of resources, especially people, to continue development and maintenance of the system.

FEATURES

A properly structured computerized work order system should do all or most of the following:

Track labor and material costs.

Produce a report of selected deferred work orders.

Provide ability to obtain work order backlog by craft code.

Produce a report of active work orders.

Produce a report of overdue work orders.

Have the ability to select work orders by any status code or a combination of codes.

Maintain an on-line, detailed work order history by equipment for at least a year.

Automatically flag corrective work orders that are for equipment under warranty.

Permit each work order to be assigned a status code which may be updated during the life of the active work order.

Assign job priorities so the most important work gets done first.

Make it possible to plan and schedule work.

Record use of special tools and materials.

Craft assignments.

HOW TO START A COMPUTERIZED WORK ORDER PROGRAM

Collecting Information

The work order number is the collection point for all information in the system. To be effective, the system requires use of written work orders with a maintenance work order (MWO) number assigned to each. The MWO number can be manually entered into the computer, or some software packages allow automatic generation of this number. The system also incorporates time and materials reporting, both made against individual work order numbers.

Capturing the desired cost and equipment history information on all work performed by the maintenance department is impossible without the consistent use of written work order requests (and entering all these requests into the computer) for both planned and unplanned jobs.

When there is more unplanned than planned work, the efficiency of the maintenance department will be low. Maintenance customers must be shown that anticipating maintenance work, and requesting a job be done before it becomes an emergency, is much more efficient and cost effective for them as well as for the maintenance group. Of course, a good preventive maintenance program also helps to avoid downtime and emergencies, as discussed in Chapter 3.

Planned work is possible when requests are received in time for planning and scheduling the labor, material, and tools needed to complete the job when required. Unplanned work is initiated by verbal requests made in response to emergencies that do not allow time for planning. Unplanned work may also be a short-duration job completed by a craftsman while assigned to other work in the same area. For best results, nonemergency jobs that require more than 1 hour of work should fall into the planned-work category, although this determination varies from industry to industry.

Setting Job Priorities

A priority rating system brings objectivity and professionalism into the planning and scheduling of maintenance work. Using a rating system does not eliminate the need for applying individual judgment in assigning final priorities, but it should eliminate the possibility of the department head with the loudest or meanest expression getting his job done first, or the likelihood of a maintenance supervisor's buddy in production getting his work done first.

The customer should always assign a priority rating to every job request. Before a work order is authorized and planned, the priority rating may be reassessed and possibly changed according to set priority guidelines.

Priorities generally are established by using either a multiplier system or a straight numeric system. Probably the best known multiplier system is the relative importance of maintenance expenditures (RIME) system, originally developed by Albert Ramond & Associates. RIME uses two rating scales, one to rate machinery, buildings, or units of equipment, and the other to rate classifications of maintenance work. Both scales are from 1 to 10, with 10 given the highest priority. On the work classification scale, maintenance, construction, and repair work that is most critical from a breakdown or safety standpoint is rated a 10 and housekeeping chores are rated a 1. On the machinery scale, utilities are rated a 10 and buildings, roads, and offices not directly influencing production are rated a 1. The highest rating is always given to work most important to production. When the machine code and the work classification code for a particular job are multiplied together, the resulting index number (such as 9 x 6 = 54) is used to schedule the job. The highest numbers are scheduled first.

Another multiplier system, developed by Patton Consultants, Inc., establishes that priority = need urgency x customer rank x equipment criticality (NUCREC). Patton believes that the RIME system is improved by adding a customer rank rating. Also, NUCREC makes 1 the top priority instead of 10. With NUCREC, the ratings included in each scale can vary from one organization to another depending on their needs. Each of the three NUCREC scales range from 1 to 4. For need urgency, the scale is: 1, emergency; 2, downtime; 3, routine work and preventive maintenance; and 4 convenience.

The customer rankings may be: 1, top management, 2, production line with direct revenue implications; 3, middle management, research and development, and, perhaps, frequent customers; and 4, all others.

The equipment criticality scale is: 1, utilities and safety systems; 2, key facility equipment with no backup; 3, impact on morale or productivity; and 4, little use or effect on output.

In the NUCREC system, a job rated 1 x 1 x 1 would get done immediately. With this system, the initial rating given a job may change with time. For instance, a PM job may be raised from a 3 to a 2 on the need urgency scale, if the job is not completed in the week originally scheduled.

A third multiplier system has been developed by maintenance management at Eli Lily & Company. This system introduces an aging factor to the process of assigning priority using an equipment value scale and a situational value scale.

The system uses a 1-10 scale for equipment value (or relative importance) and a scale of 1-4 for situational value, such as a safety hazard or critical production breakdown, either of which would be rated a 4. In this system, the highest numbers receive highest priority. The initial priority assigned to a job is the equipment value multiplied by situational value. Therefore, if the equipment value is 7 and the situational value is 2, the initial priority rating is 14. The aging factor in the system provides that every priority moves from its initial value to 80, which equals the due or start date. Emergencies age to 80 immediately, making them due immediately. For the system to work, the requested date and due date must be entered on the work order.

If the due date is given for 11 days later and the initial priority is 14, then each day the priority value would increase by (80 - 14)/11, or 6, so that in 11 days the priority would have reached 80. The aging factor helps the planner to sequence backlogged work. Due dates, of course, can be changed as new work orders come in. Planners may use 90-99 to handle emergencies. Other modifications of the basic multiplier system can be devised to reflect a maintenance organization's particular needs and philosophy.

Priority definitions for a straight numeric system of 1-5 with 1 representing highest priority may be as follows:

1. When an extremely unsafe condition exists that may endanger life or limb, damage equipment, or contaminate

the environment, or critical production equipment is down. In such cases, overtime will be assigned, if necessary, to handle the repair work.

2. Urgent work that should be started as soon as possible, usually within 24 hr. A safety or equipment hazard will exist or production will be down unless repairs are made. This rating may also be assigned to planned shutdown work that is necessary before equipment can be started.
3. Essential work for which immediate completion is not critical, but work should be started within the next week if parts, materials, and equipment are available. This category may include preventive maintenance and minor modifications and small jobs that can be assigned as fill-in work.
4. Work to be scheduled when parts, materials, equipment, and personnel are available. This work is required but can be used to accumulate backlog, because it is not urgent.
5. Work requiring extensive planning and scheduling that may be designated for a shutdown period. This category may also include jobs that can be delayed for several months and used as fill-in. These jobs may require engineering or design work and therefore need more lead time and coordination with production, maintenance, and vendors or contractors for completion at a scheduled time.

If a priority system is part of a computerized maintenance system, the planner must have the option of overriding the computer recommendation. A change in the rating may be necessary because of unusual circumstances, such as an approaching shutdown, or other knowledge the planner has that the computer may not have.

Designing the Work Order

There are many variations in printed work order forms (Figure 4.1). Work order design and contents should be considered care-

```
WORK ORDER NUMBER: 0822850001   ! PRIORITY: 05   ! WORK CAT: BK
--------------------------------------------------------------------
 DATE ISSUED: 08/22/85  14:30      REQUESTED BY: RJK   PHONE: 782-5130
 DATE REQUIRED: 08/24/85  00:00     APPROVED BY: CDS
 ACCOUNT #: LM2212                 EQUIPMENT NO: U4473
 EQUIP. NAME: PRESSURE TANK
 EQUIP. DEPT: MFG
 EQUIP. LOC.: BLDG 445 R4P17
 FAILURE CODE: VF                       JOB NO:
 REMARK:                                WORK REQUEST NO: TK445
--------------------------------------------------------------------
 LABOR:      CRAFT      REG HRS     OVERTIME HRS     TECHNICIAN
             MECH
--------------------------------------------------------------------
WORK DESCRIPTION
High pressure tank emergency release valve is not testing properly.

Valve must be repaired or replaced before line 4 resumes operation
on Monday.
--------------------------------------------------------------------
PART NUMBER    DESCRIPTION            LOCATION      QTY    ACT QTY
01-03-44          VALVE REPAIR KIT    Q3S4           1
0159182-00        RETAINING RING      O5S1           4
--------------------------------------------------------------------
 DATE COMPLETED:        !  OUTSIDE COSTS:        !  CHECKED BY:
                        !                        !
--------------------------------------------------------------------
 CURRENT EQUIP          !  LOST        HOURS:         COST:
       USAGE:           !  PRODUCTION
--------------------------------------------------------------------
  ACTION     1.            2.           3.           4.
  CODES
--------------------------------------------------------------------
```

Figure 4.1 Typical work order form. (Courtesy of UNIK Associates.)

fully, according to actual need and use. The work order must be easy to use. Although its day-to-day application should supply all the information maintenance requires, the work order should not ask for excessive or extraneous data. The essential information on a work order defines the job and provides for planning, scheduling, control, and accumulation of equipment history (Figure 4.2).

In computerized systems a single-copy printout of the work order can be given to the craftsman. Everyone else can obtain desired information by using a CRT terminal. [In a manual sys-

Purpose	Information requested
To define the job	• A unique job number • Equipment number, description, and location • Account number • Budget/project number • Priority code • Type of work code • Work requested
Planning, scheduling, and control	• Labor estimate • Material estimate • Job plan • Total estimated cost • Approvals
For control and history	• Actual labor costs • Actual material costs • Actual work done • Failure codes

Figure 4.2 Work order system design. (Courtesy of Arthur S. Dearden, The Emerson Consultants, Inc.)

tem, a multicopy form (four or more) is used. Each copy is a different color and is designated for use by different people: the originator, planner, supervisor, craftsman, etc.]

Using the Work Order

Everyone who uses a work order form must be instructed in its use. In most plants, written procedures are needed that detail use of the entire work order system. The procedures should outline the responsibilities of the customer and of maintenance. Accountability is needed on both sides.

In most cases the work order flow is from the originator, to the maintenance clerk or planner, to the maintenance super-

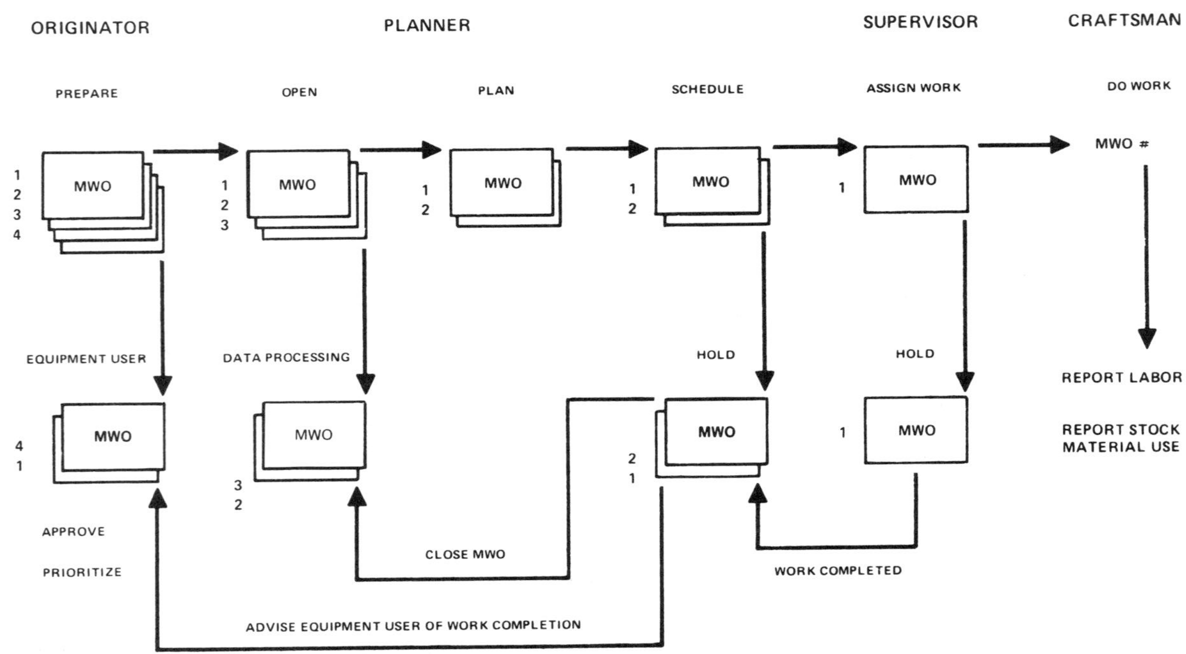

Figure 4.3 Utilization of maintenance work order (MWO). (Courtesy of Paul D. Tomlingson Associates, Inc.)

visor, to the craftsman, and back to the planner for closing (Figure 4.3).

Each of these individuals must know what information he is responsible to provide on the work order so that it can be entered in the computer system. Generally, the system works in the following way:

1. A central office and telephone number are used to receive all maintenance work requests and work orders, because control is lost when requests are made directly to supervisors or craftsmen.

2. The work order originator (or customer) identifies himself, the equipment or system that needs attention and its location, and provides a clear and complete description of the problem and the work to be done. No more than one job is entered on a work request or work order form. A reason is given for the request. Some work order forms also ask the originator to identify the type of maintenance work being requested by using a code that classifies a job as preventive maintenance, corrective repair or replacement, shutdown repair, minor modification, and so on.

Equipment is identified by a number that can be the nameplate or tag number given to the equipment when it came into the plant. If equipment does not have tag numbers, maintenance establishes a system for assigning numbers.

The cost center is also identified by number. In addition, an account or subaccount number may be used to identify the department represented. These cost center numbers and the equipment numbers allow maintenance to charge costs against both the customer and the particular piece of equipment.

The work order may also require the customer to indicate the crafts and materials needed to do the job.

The customer assigns a priority rating to the work request according to the established maintenance priority system. A brief description of the priority code can be printed on the work order for reference.

Finally, the originator enters the date of the request and the desired completion date for the job. He keeps one copy of the work order and forwards the others to the maintenance office.

The maintenance planner receiving the work order reviews the job description and may visit the work location to ascertain what is required to perform the work. If his estimate of labor hours and materials for the job differs significantly from the customer's, the planner will contact the customer for final authorization before planning and scheduling the job. The priority rating may also be discussed at this time.

Until the proper approval and authorization is received for a job, the work order is only a request. A system for authorization and approvals of maintenance work should be established by plant or corporate management. The maintenance customer should be responsible for adhering to the rules.

The planner then, with the help of the computer program, schedules the job and makes sure that labor, parts, and materials are on hand, and the equipment is shut down or otherwise available for work.

3. The maintenance supervisor reviews the work orders with the planner. In some organizations, a maintenance clerk accepts orders and routes them to the maintenance supervisor, who does the planning, scheduling, and assigning of work. The supervisor gives the work order to the craftsman assigned to the job.

4. The craftsman, on completion of a job, adds to the work order information that describes the actual work performed, materials used, and hours worked. The craftsman may also fill in failure analysis information. When a craftsman finishes a job, he should sign the work order and return it to his supervisor, who then verifies the work description, materials used, and hours spent. The supervisor returns the work order to the maintenance planner or clerk, who enters this additional information into the computer.

Time and Material Reporting

Time: Craftsmen must be responsible for reporting their time accurately and by work order number. A good system includes reporting of delay time (waiting for an assignment, tools and materials, or release of equipment), as well as the time spent working on each job. Upon completion of the job, this information is entered into the computer.

Material: Other essential information connected with the work order system is a record of the spare parts and materials used to complete a job and their actual cost. The program contains a datafile that shares all the parts in stock and their unit cost. When the work order is issued, a list of required parts and material is prepared and entered into the computer. The printed work order that goes to the craftsman shows this list of materials. The craftsman should be instructed to check the list O.K. if he uses all the parts and material as specified on the work order. If it differs from the specified list, he should record it on the work order. Again, upon completion of the job, this information is also entered into the computer.

REPORTS

Maintenance costs can be reduced drastically if equipment failures are minimized. The equipment history report, as shown in Figure 4.4, includes the number of work orders charged to each piece of equipment, the costs of labor and materials, and the total main-

MO9R EQUIPMENT MAINTENANCE HISTORY FOR THE PAST 4 QUARTERS, AS OF 1/29/78
PAGE 3

EQUIPMENT NO.	NO. OF WOS	COST OF MAINT LABOR	COST OF MATERIAL	TOTAL MAINT COST	REPLACMT COST	PCT. MAINT. COST IS OF REPL. COST
10002	1	43.13	0.00	43.13	4000.00	1
16025	5	34.38	12.00	46.38	200.00	23
16050	2	31.25	30.00	61.25	200.00	31
10070	3	45.31	30.80	82.11	2500.00	0
16100	6	117.19	135.00	252.19	400.00	63
16150	2	25.00	200.00	225.00	630.00	36
16175	18	850.00	375.00	1225.00	1052.00	116

HIGH REPAIR COST. CHECK CAUSE.

Figure 4.4 Equipment maintenance history. (From T. M. Hershey and F. Claire, *Plant Engineering*, July 26, 1979, p. 97.)

M10R WORK ORDERS COMPLETED FOR SELECTED EQUIPMENT, 1/29/78
PAGE 1

EQUIPMENT NO.	WO. NO.	DATE WO CLOSED	JOB DESCRIPTION	ACTUAL HOURS	MATERIAL COST
16175	4402	1/16/78	NO. 16175 PUMP REPAIR	5.0	150.00
	3979	1/15/78	REINFORCE PUMP BASE	0.5	0.00
	4301	1/13/78	REPL SEAL	11.0	150.00
	3271	12/9/77	CHECK PUMP	1.5	0.00
	3459	12/9/77	REPLACE SEAL	10.5	150.00
	2678	10/8/77	REPLACE SEAL	0.3	150.00
	2980	9/30/77	PUT NEW SEALS IN PUMP	12.0	150.00
	2413	9/4/77	CHECK PUMP	4.5	0.00
	2414	9/4/77	REPL SEALS	12.0	150.00
	2415	9/4/77	REPL GASKETS & BOLTS	2.0	35.00
	2228	9/2/77	CHECK PUMP	4.0	5.00
	2404	8/31/77	CHECK REPORTED PUMP NOISE	8.0	40.00
	1949	5/15/77	REPAIR SEALS	12.5	150.00
	1276	5/8/77	PAINT PUMP PLATFORM	9.8	10.00
	1138	3/7/77	REPL SEALS	12.5	150.00
	1007	2/8/77	ADJUST PUMP	1.0	0.00
	976	2/8/77	CHECK PUMP	0.5	0.00
	732	1/30/77	CHECK PUMP OPERATION	0.5	0.00

Figure 4.5 Work order for selected equipment. (From T. M. Hershey and F. Claire, *Plant Engineering*, July 26, 1979, p. 97.)

tenance cost for the past year. It also shows equipment replacement cost and maintenance cost as percentage of replacement cost. This report helps management determine which ongoing maintenance costs are acceptable.

The report in Figure 4.5 provides additional information on equipment with high maintenance costs. This report gives

M07R BACKLOG SUMMARY FOR THE DEPARTMENT, WEEK ENDING 1/29/78

	NUMBER OF MAN-HOURS PLD. OR EST.			
CRAFT CODE	OPENED	CLOSED	CANCELLED	BACKLOG
ELECT.	5.8	9.0	0.0	12.7
WELD.	160.0	160.0	0.0	166.4
MECH.	435.0	496.4	0.0	1808.0
TOTAL	600.8	665.4	0.0	1987.1

INCLUDES CARRY OVER BACKLOG HOURS FROM PREVIOUS WEEKS, NOT REPORTED SEPARATELY.

Figure 4.6 Backlog summary report. (From T. M. Hershey and F. Claire, *Plant Engineering*, July 26, 1979, p. 97.)

work order numbers, descriptions of work, labor hours, and material costs. The job descriptions pinpoint the seals as this pump's primary maintenance problem. Investigation should reveal why the seals have short lives and what corrective action should be taken.

The computer can supply additional reports that are important to the plant engineering department. Some of these reports are work backlog, work order status, and work orders past due.

A knowledge of the backlog is necessary to plan crew sizes, schedule overtime, and determine the need for hiring a subcontractor. The backlog summary (Figure 4.6) shows the number of man-hours planned or estimated on jobs that have been opened, closed, and canceled for each craft during the week, and total backlog hours. Figure 4.6 shows that more jobs were closed than opened, indicating that the department reduced its backlog during the week.

The breakdown by craft is particularly helpful. In the example, there is almost no work backlog for the electrical craftsmen, but a great deal for the mechanics. The maintenance department should give this area careful attention, because it is usually difficult to transfer manpower from one craft to another.

Similar reports for each maintenance crew provide supervisors with additional detail for planning.

The maintenance manager should occasionally request summarized information on work orders for major or critical jobs. Figure 4.7 lists work orders selected for monitoring. Note that the actual hours on many of the work orders substantially exceed the hours planned or estimated. Jobs showing excessive hours and material costs, or little progress, should be watched closely and delays should be investigated.

Another administrative maintenance report lists all past-due work orders. Figure 4.8 shows work orders that have been open for 30 days or more. Regular issuance of this report keeps managers from forgetting overdue work orders, particularly those that have been deferred because of low priority.

M11R STATUS OF SELECTED WORK ORDERS, AS OF 1/29/78
PAGE 1

WO NO.	AREA	JOB DESCRIPTION	EQUIPMENT NO.	SUPV	PRIORITY CODE	PLD/ EST.	HOURS PLD OR EST.	ACTUAL HOURS	MATERIAL COST	DATE OPENED	DATE CLOSED
24	12	GRINDING MURRAY CHIPPER KNIFES	12250	2	SW	EST	41.0	401.5	0.00	4/6/77	
28	48	UNION + TRAINING MEETINGS	383	2	SW	EST	66.0	747.2	0.00	6/16/77	
646	30	INSTALL ENGINE HEATER	46163	2	HP	EST	6.0	5.0	125.00	11/19/77	12/7/77
647	48	INSTALL SAFETY POSTER FRAMES	383	2	LP	PLD	1.3	2.5	35.00	12/8/77	12/9/77
714	26	CHG RNNR W W PUMP TO PAPER M	32111	2	SD	EST	4.0	19.0	385.50	4/6/77	4/7/77
717	44	INSTALL DOOR CLOSURE	387	2	LP	EST	1.0	1.0	4.45	4/11/77	4/11/77
720	26	PIPE AIR TO RECEIVING TANK	301	2	HP	EST					
743	12	INSTALL GUARD SLAT CONV MTR	21768	2	HP	EST					
750	44	RPR STEPS N SIDE RFF BLDG	387	2	LP	EST					
760	14	CHG BACK STUFFING BOX WW PP	32113	2	EM	EST					
761	26	RPR DRYER JOINTS	301	2	SD	EST					
762	22	RPR HYDRAPULPER SEAL	22000	2	SD	EST					
1602	24	RPL DOCTOR BLADE HOLDER 6A	24025	2	LP	EST					

NOTE:

PRIORITY CODE	MEANING
EM	EMERGENCY
HP	HIGH PRIORITY
LP	LOW PRIORITY
SD	SHUT DOWN
SW	STANDING WORK ORDER

Figure 4.7 Status of selected work orders. (From T. M. Hershey and F. Claire, *Plant Engineering*, July 26, 1979, p. 98.)

M12R WORK ORDERS OPEN FOR 30 DAYS OR MORE, AS OF 1/29/78
PAGE 1

AREA	DATE	WO NO.	DESCRIPTION	PRIORITY CODE	EQUIPMENT NO.	SUPV	NUMBER OF MAN-HOURS PLD.	EST.	ACTUAL	DAYS OPEN
20	4/5/77	603	INSTALL 48 INCH HYDRASIEVE	HP	311	2		52.0	0.0	300
20	4/17/77	311	RPR FEED CONTROL ROTARIES	EM	311	4		1.0	3.0	288
20	7/29/77	1839	CHG PLATES NO. 2 REFINER	LP	20078	2	4.8		0.0	185
20	10/2/77	3184	MACHINE FLYTES PRESSIFINER	LP	311	2		6.0	35.5	120
20	12/2/77	3039	INSTALL LINERS 410 REFINERS	LP	383	2	8.2		0.0	59
20	12/19/77	3879	CK DRIVE ROSS SAVEALL	LP						
20	12/20/77	3887	RPR TELEPHONE REFINER	LP						
20	12/27/77	3906	CHG PLATES SPROUT REFINER	SD						
20	12/27/77	3909	CK BROKE ADJ DRIVE	HP						
20	12/18/77	4225	CLN & CHECK SPROUT CONTROLS	SD						

20 — 11 WORK ORDERS ARE OPEN

NOTE:

PRIORITY CODE	MEANING
EM	EMERGENCY
HP	HIGH PRIORITY
LP	LOW PRIORITY
SD	SHUT DOWN
SW	STANDING WORK ORDER

Figure 4.8 Work orders open for 30 days or more. (From T. M. Hershey and F. Claire, *Plant Engineering*, July 26, 1979, p. 98.)

```
           WORK ORDER MATERIAL REPORT          01/24/85     PAGE 1

ALL RECORDS SELECTED

WORK ORDER NO.   QTY PART NUMBER   DESCRIPTION               UNIT COST  EXT COST
--------------   --------------    -----------               ---------  --------
0123850001       *** NO MATERIALS ***

0123850002       *** NO MATERIALS ***

0123850003       *** NO MATERIALS ***

0123850004         5  AN200          1/4" X 2" X 2" ANGLE        17.50      87.5
                   2  BT2043         CRANK CLAMPING BOLT          5.95      11.9
                   5  W003                 *** ITEM NOT IN INVENTORY ***
                                                                     -----------
                                                                            99.4

0123850005        10  AN200          1/4" X 2" X 2" ANGLE        17.50     175.0
                  10  BT2043         CRANK CLAMPING BOLT          5.95      59.5
                                                                     -----------
                                                                           234.5

0123850006       *** NO MATERIALS ***

0123850007         1  B204           V-BELT                       9.50       9.5
                                                                     -----------
                                                                             9.5

0123850008         4  AN200          1/4" X 2" X 2" ANGLE        17.50      70.0
                   3  S7                   *** ITEM NOT IN INVENTORY ***
                   4  W003                 *** ITEM NOT IN INVENTORY ***
                  15  RD90                 *** ITEM NOT IN INVENTORY ***
                                                                     -----------
                                                                            70.0

                                                                     ===========
                                                                           413.4
```

Figure 4.9 Work order material report. (Courtesy of UNIK Associates.)

Work orders 603 and 311 have high priority and emergency priority, respectively, but they have been open for 10 months. This situation should be investigated and corrected. An error in paperwork could be responsible for keeping these jobs in the computer files.

Work order 3184 shows 6.0 estimated hours and 35.5 actual hours; it has been open for 4 months. Investigation may reveal that there is a problem in performing the work, or that craftsmen are improperly charging items to this work order number. All jobs listed should be reviewed to guarantee that they are scheduled and not overlooked.

WORK ORDER LABOR REPORT 01/28/85 PAGE 1

ALL RECORDS SELECTED

WORK ORDER	CRAFT	TECH	EST HRS	ACT HRS	OT HRS	COST	%PROD
0123850001	MECH		1.00	1.00	0.50	24.50	66.67
			1.00	1.00	0.50	24.50	
0123850002	MECH	MX	0.25	1.00	0.75	29.75	14.29
			0.25	1.00	0.75	29.75	
0123850003	PIPE	ER	1.00	1.00	1.00	30.00	50.00
			1.00	1.00	1.00	30.00	
0123850004	MACH	LK	6.00	8.00	0.00	96.00	75.00
	WELD		4.00	0.00	0.00	0.00	
			10.00	8.00	0.00	96.00	
0123850005	WELD	PO	16.00	20.00	4.00	338.00	66.67
			16.00	20.00	4.00	338.00	
0123850006	ELEC	YU	1.50	2.00	0.00	28.00	75.00
			1.50	2.00	0.00	28.00	
			29.75	33.00	6.25	546.25	

Figure 4.10 Work order labor report. (Courtesy of UNIK Associates.)

WORK ORDER COST SUMMARY 01/24/85 PAGE 1

ALL RECORDS SELECTED

WORK ORDER NO.	LABOR COST	MATERIAL COST	OUTSIDE COST	TOTAL COST
0123850001	24.50	0.00	0.00	24.50
0123850002	29.75	0.00	0.00	29.75
0123850003	30.00	0.00	0.00	30.00
0123850004	96.00	99.40	0.00	195.40
0123850005	338.00	234.50	0.00	572.50
0123850006	28.00	0.00	0.00	28.00
0123850007	0.00	9.50	0.00	9.50
0123850008	0.00	70.00	0.00	70.00
	546.25	413.40	0.00	959.65
	546.25	413.40	0.00	959.65

Figure 4.11 Work order cost summary. (Courtesy of UNIK Associates.)

```
          ACTIVE WORK ORDER REPORT                       01/23/85      PAGE 1
ALL RECORDS SELECTED

WO NUMBER    DATE DUE    PRTY  CRAFT   EST HRS   REQ BY  PHONE     EQ NUMBER
---------    --------    ----  -----   -------   ------  -----     ---------

0123850001   01/20/85     P2   MECH      1.00                      603
Check oil, brakes, tires, charging, ignition system and horn.

0123850002   01/20/85     P2   MECH      0.25                      A-11
Check ball bearing drive, screw and nut for alignment, pitting, scoring,

0123850003   01/20/85     P2   PIPE      1.00                      P423
Check pressure on hydraulic system.

0123850004   02/18/85          MACH      6.00      ES    4795
Fabricate fixture and tool for turret lathe

0123850005   02/17/85          WELD     16.00      SP    4793
Build storage rack per drawing and instructions.

0123850006   01/30/85          ELEC      1.50      JD    4791      W751
Replace time relay.

0123850007   01/25/85          MACH      1.00      JS    4789      B4321
Check belt and pulley.

0123850008   02/18/85          MACH      6.00      ES    4795
Fabricate the die for offset press.
       END OF REPORT
```

Figure 4.12 Active work order report. (Courtesy of UNIK Associates.)

Figure 4.9 shows a work order material report that displays quantity, part number, description, and cost for each work order. The work order labor report (Figure 4.10) shows craft, description, estimated and actual hours, labor cost, and productivity for each work order. Note that productivity (%) is computed based on estimated and actual hours.

Figure 4.11 shows a work order cost summary that includes work order number, labor cost, material cost, and total cost. This is a very important report for the maintenance manager. The same report can be obtained for all repair jobs or for any other specified work category.

Figure 4.12 shows all work orders that are being processed.

CONCLUSION

A properly controlled work order system allows the maintenance manager to obtain the information needed to analyze costs by job, units of equipment, priorities, crafts, cost centers, and account numbers. Data are generated to monitor work order status and backlogs, add to or start equipment histories, and control manning levels.

A computerized system will not work at an optimum level unless management is fully committed to its use and all appropriate employees are trained in the procedures. When a new system is developed, input from all users should be requested and fully considered. A team effort is required for success.

A computerized work order system provides needed feedback on maintenance performance and costs. This information can be used periodically to change and upgrade the system, whether this means making organizational changes, streamlining PM, or improving materials control.

5

Spare-Parts Inventory Control

INTRODUCTION

Maintenance cannot keep machines running and facilities in good condition if the company does not provide a good spare-parts inventory system. A spare-parts system should stretch the parts budget, help to reduce machine downtime, and provide a means of communicating with, and educating, the maintenance work force. And, it should provide pertinent data for developing work order costs and machine history records. Materials typically represent 50% of the total cost of maintenance. Even with this high percentage level, the importance of properly managing this valuable asset is overlooked in most organizations. Industry can realize significant savings by controlling inventories of maintenance materials. Management is becoming more aware of the important part a well-maintained storeroom can play in reducing breakdowns and eliminating plant shutdowns. This chapter will offer proven methods for better managing materials with full utilization of the computer.

MATERIAL CLASSIFICATIONS

The first step in organizing an inventory program is to classify materials according to importance:

Critical insurance parts: Spare parts that are critical to the operation of essential machines (parts that cannot be used elsewhere). The downtime cost of not having the part usually exceeds the value of the part. Because usage of these insurance parts is unpredictable, a spare must always be in inventory.

Standard replacement parts: Parts that can be used on more than one machine or piece of equipment. Some are used frequently and have a high unit cost, some may be low in usage and cost, and others may be combinations of high, medium, and low usages and costs.

Parts and supplies purchased for particular jobs: These materials go directly to a storage area for the specific job or into stores inventory and add to current quantities until needed. After the job has been completed, surplus materials go into inventory or are scrapped.

Hardware items: These include bolts, nuts, washers, cotter pins, and so on.

General supplies: Used in maintenance and repairs.

Tools: Defined as items that are stored with the usual maintenance material to be checked out by the maintenance personnel for a specific work order task. It is necessary for tools to be included in the maintenance system, since maintenance planning and scheduling must include those items.

UNDERSTANDING INVENTORY TERMS

The following is a list of the most common inventory terms, with explanations of how they fit into an inventory control system.

Annual usage: The number of items withdrawn from stores over a period of 1 year. The plant engineer needs to know the exact amount so he can decide when to reorder, allowing time to order, receive, and place in storeroom. Possible fluctuations in usage must also be analyzed. The more stable the use rate, the less likelihood of a stockout.

Unit cost: The price of a particular item. A formula to decide how much should be ordered has to include the unit cost.

Lead time: The time it takes to obtain an item. Lead time includes the time to process an order, the time to ship it, and the time to stock it. Lead time is not, as commonly believed, the time from issuance of the purchase order to receipt of goods.

Order cost: The cost of issuing a purchase order for a certain quantity. It includes the cost of labor and overhead, paper, office machines, and postage in all departments that go through the procedures of issuing the requisition, receiving the material, and delivering it to the storeroom.

Carrying cost: The cost of storing an item in the storeroom. It includes the charge for floor space, depreciation, taxes, interest, storeroom operating charges, etc.

Space: The cost of storage space is determined in a variety of ways. The cost of constructing the storehouse may be debited according to the square footage or cubic footage used for storage of material, and this cost might be depreciated over a period of time.

Special conditions: Many items require a specialized atmosphere—air conditioning, humidity control, heat, or ventilation—in order to prevent their deterioration.

Operating cost: Operation of the stores function involves cost of maintaining files and computer programs for all items in stock.

Materials handling: The cost of the entire system—often from the supplier's factory to the point of usage in the plant.

Obsolescence and the like: Some portion of stored material will become obsolete, misplaced, lost, or pilfered; or the equipment for which spares were purchased may be scrapped —all of which adds to the cost of storage. Many items have a finite shelf life, and deterioration adds to the other costs.

Taxes and insurance: The average value of the storehouse is used to compute both taxes and insurance.

Interest: Interest, or cost of money, is probably the largest of these charges.

Safety stock level: The stock of an item on hand for protection against running out when delivery is delayed or use is higher than normal. It can be thought of as insurance. To determine the quantity of safety stock, the plant engineer must consider the cost of not having the item in terms of lost production and and lowered craft productivity versus the cost of carrying the added inventory. The amount of material depends on lead time for that item, its use rate, and its importance to the production effort.

If the lead time is very short, the material is readily available from several sources, and it is not critical to operations, you might opt for a zero safety stock level. If the lead time is longer, there are only a few alternative sources, and the material is of some degree of criticality, the level should be set at some minimum figure. If the lead time is relatively long, the material has a single source, and it is critical to plant operations, a maximum safety stock level should be established.

Expedite point: The point in the inventory level at which the safety stock is used. At this point, it is important to follow up on an order to be sure it is received before zero inventory is reached. If the expedite point is zero, the item is not critical to the production effort and no safety stock is really needed.

Reorder level: The low limit for the sum of the in-stock and on-order balances that is used to determine when to order another lot. The reorder level is determined from data on use and lead time. It is the lead time in months multiplied by usage per month plus the safety stock level.

Reorder level = lead time (months) x usage per month
+ safety stock level

Economic order quantity: The quantity that will result in the lowest total costs of ordering, making, or procuring the item, and carrying the resulting inventories. The theory of economical order quantity buying is that the cost of carrying inventory and the cost of buying are like balance scales: when one goes up, the other goes down.

Calculating the economic order quantity: The following formula can be used to determine the quantity of an item that will result in the lowest total cost of reordering it and carrying an inventory:

$$Q_E = \sqrt{\frac{2SP}{IC}}$$

where

Q = economic order quantity, units
S = annual usage, units
P = cost of issuing purchase order, dollars
I = inventory carrying cost, percentage
C = unit cost of item, dollars

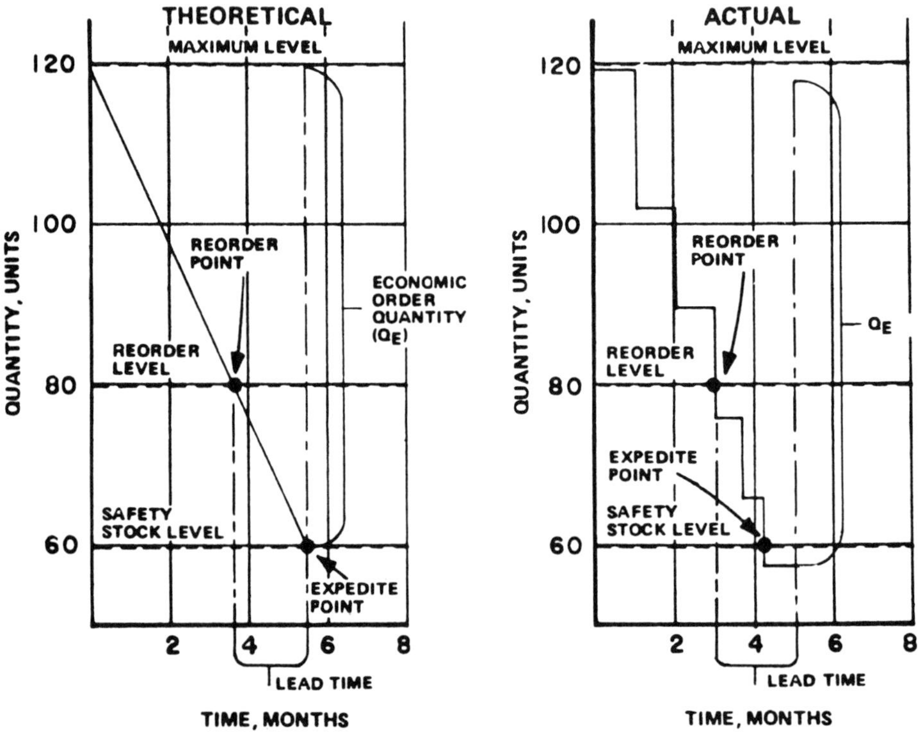

Figure 5.1 Safety and cycle stock. (From W. H. Weiss, *Plant Engineering*, July 8, 1982, p. 59.)

The reasoning behind this formula, along with the way the curves were developed, is shown in Figure 5.1.

To illustrate the formula, a \$50 item that the plant uses at the rate of 75 a year can be considered. The cost of issuing a purchase order runs about \$65, and the inventory carrying cost in industrial plants is about 30% of the value of storeroom inventory. The economic order quantity would be 26.

To show the order quantity range possible in a plant, four examples are shown, using the same ordering cost and carrying cost.

Annual usage	Unit cost, dollars	Q_E
5	10	15 (14.7)
5	3000	1 (0.85)
7000	1	1740 (1741.6)
7000	20	390 (389.4)

Following is a purchasing problem to illustrate the value of the economic order quantity formula:

You are a buyer of maintenance materials. Periodically, you order additional solenoids to replace the storeroom stock. The local vendor has been supplying solenoids at $100 each when 26 are ordered. But today, when the order is placed, he offers them at $90 each if 50 are ordered. That is a 10% discount, but is it a good buy? How much would you actually save the company if the offer were accepted?

To solve this problem, use the equation that applies to inventory control:

$$T = \frac{Q}{2}IC + \frac{SP}{Q}$$

where

T = total cost, dollars

Q = quantity ordered, units

Records show that the plant uses 75 of the solenoids in a year. It costs $65 to process a purchase order, and the inventory carrying cost is 30% of the value of the storeroom inventory.

Let T_1 = total cost of usual order and T_2 = total cost of today's offer.

$$T_1 = \frac{26}{2} \times 0.3 \times 100 + \frac{75 \times 65}{26} = \$577.5$$

$$T_2 = \frac{50}{2} \times 0.3 \times 90 + \frac{75 \times 65}{50} = \$772.5$$

Because T_2 is greater than T_1, the vendor's offer would cost the company more. It should be turned down.

> *Maximum level*: The sum of the stock on hand at the safety level (expedite point) plus the order quantity.
>
> *Stockout cost*: The cost of a stockout can be only a few dollars, or thousands of dollars if a production line or the plant goes down. A part may be ordered and received before it is needed again. In this case the only extra cost is the expediting cost to get the part into stock before it is needed.

The risk of stockouts is greater for parts with longer lead times. Parts that are used frequently may be replenished faster than those that are used less often or that are unique to certain machines or pieces of equipment. The cost impact of a stockout usually declines as the operation moves from the beginning to the end of the production process. For example, in the food industry the breakdown of a large mixing machine will have more impact on cost than the breakdown of a packaging machine. Some of these terms are shown in Figure 5.2.

CONTROLLING INVENTORY

The best means of controlling inventory is to have the right quantity in the right mix and at the lowest cost.

As materials are used over a period of time, a reorder point is reached and a requisition is issued to replenish inventory. The material should be received before a stockout occurs. Including a safety stock provides a cushion to help prevent a stockout.

Inventory control does not necessarily mean that inventories should be kept at the minimum. The lowest possible inventory is

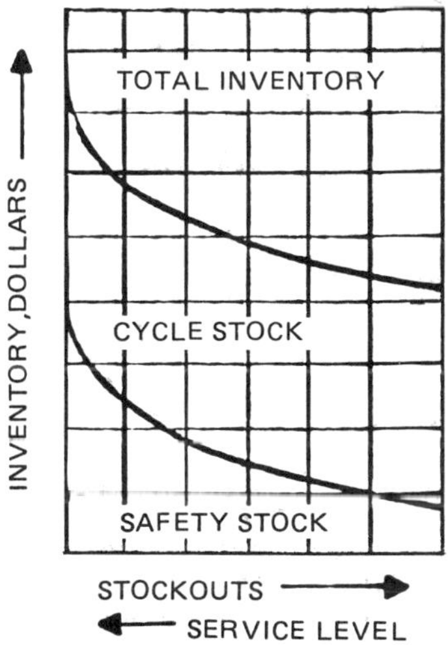

Figure 5.2 Terms mentioned in the text. (From W. H. Weiss, *Plant Engineering*, July 8, 1982, p. 59.)

often not the best or least costly in terms of total cost. Although the relative importance of various inventory control objectives may change with changes in business conditions, the need for effective control of inventories is constant.

Inventory control is best approached by working out answers to these questions:

What items need to be ordered?

When should an item be ordered?

How much of it should be ordered?

From whom should it be ordered?

What to Stock

When a plant is initiating a storehouse or drastically revising a storehouse system, it is necessary to have a systematic way of determining what material should be stocked.

Spare parts are usually stocked in accordance with the manufacturer's recommendation and in light of plant experience. If the plant is new and does not have any experience, the manufacturer's list of recommended spare parts should be considered.

How Much to Order

Each of the following categories should be considered separately in deciding which method to employ.

Critical "Insurance" Parts: The economic order quantity formula and safety stock calculations cannot be used for these parts. Usually, there are only one or two spares. With two spares, a replacement should be ordered routinely when one of the spares is taken from inventory.

Insurance parts should be tightly controlled. Access to these parts should be limited. If a part withdrawal is not reported, parts may not be available when needed.

These parts usually represent a large dollar percentage of inventory. Inventory can be limited in two ways: by not ordering more than the maximum and by reducing the maximum from three items to two or from two to one.

Standard Replacement Parts: These are ideal candidates for economic order quantities and safety stocks. Systems contracts and blanket orders can be set up with suppliers for more frequently used parts. Prices and paperwork will be reduced because the supplier will carry the inventory.

Parts and Supplies Purchased for Jobs: These parts should be ordered for specific jobs and should be staged or placed in storage by job numbers. If they go into inventory as part of the balance of an inventory item, the job quantities should be noted.

When planned maintenance is lacking, these materials can build up the inventory. The buildup results because the materials are not used on the job, or the job is changed, delayed, or postponed.

Hardware Items: Large quantities of these parts can be ordered with little dollar effect on the inventory while giving a high service level. Blanket orders or systems contracts can be isssued.

Supplies: Inventory quantities should be set so that supplies last for shorter periods than hardware items. These items can be purchased locally, eliminating the need to carry large inventories of such supplies.

SERVICE LEVEL

If materials are always available when needed, the company has a 100% service level. But because some materials may be out of stock, and others may not be carried, such a service level is difficult and costly to maintain. If 90 request out of every 100 are filled from inventory, the company has a 90% service level. Service level is a management decision. Once the level has been established, calculations can be made to determine the inventory required to support that level. There are three types of service levels:

1. *Overall service level*: the number of material requests filled directly from inventory. Calculate the number of requests filled by the storeroom divided by the total number of request.
2. *Inventory items*: the number of requests for inventory items filled directly from inventory. Calculate the number of items filled from inventory divided by the total number of requested inventory items.
3. *Material by category*: the number of requests for inventory items by category, filled from inventory. Calculate the service levels for each of the five categories of materials mentioned earlier.

The higher the service level, the greater will be the cost of providing that service. A higher service level means fewer stockouts; more stockouts reduce the costs of service. Service level is related to the number of exposures to a stockout during a time period. An exposure occurs with each reorder of materials. If 20 parts are ordered 300 times in 10 years and the number of stockouts during this period is 15, the service level is:

$$1 - \frac{15}{300} \times 100 = 95\%$$

The cost of service is the inventory investment to provide the required service level. Safety, or reserve, stock (to protect against stockouts) and the cycle stock (consumed between replenishments) are the main components of inventory investment (Figure 5.2).

LEVELS OF DEVELOPMENT OR SOPHISTICATION

There are four levels of development or sophistication. You should determine at what level you are.

Level 1. No Formal System, No Real Management Controls

The results of this level of control system are:

1. Not having the right parts when they are needed.
2. Buildup of slow-moving and obsolete inventory.
3. Dissatisfaction by maintenance with the delays and lost time due to material shortage.

Level 2. Simple Manual System, Some Management Controls

1. Simple work order system—some planning of material needs.
2. Stores department is closed and manned on the day shift.

3. Traveling requisitions may be used to ask the purchasing department to buy stock materials.
4. Predictable materials are listed on the work order. An equipment history file is maintained for each piece of critical plant equipment.
5. Continuous inventory records may be kept in stores.
6. Justification procedures for stocking spare parts, reorder points, and order quantities probably have been established.

Level 3. Computer-Assisted System, Fair to Good Management Controls

1. Predictable material is included on work orders. Material needs are communicated on a priority basis to stores.
2. Matching of requirements against inventory available and on hand in the computer.
3. Printing of material pick lists to stores.
4. Automatic reorder of standard replacement parts and supplies.
5. Reporting of actual material costs charged to specific or standing work orders.
6. Accumulation of material cost for repairs and maintenance by piece of equipment.
7. Formal justification procedures for stocking, replenishing, and making the spare parts obsolete.
8. Material catalogs by piece of equipment showing the spare parts stocked and data about the parts.
9. Cross-referencing of company and manufacturer part numbers.
10. Delivery of materials, designated drop points, and free issue items.

11. Systems contracts in effect with periodic releases.

Level 4. On-line Computer System, Good to Excellent Management Controls

1. Work orders entered on-line in real time by CRT to the computer.
2. Inventory allocation to work order jobs and adjustment to available and on-hand balances by computer.
3. Pick lists, on either a CRT or printed copy, produced in the storeroom by computer. Manual or automated AS/RS picking of materials.
4. Current inventory balance displayed on CRT.
5. Material catalogs on the CRT as well as printed.
6. On-line cross-referencing of parts and equipment.
7. Performance results calculated by the computer system.

SCHEDULING JOBS BASED ON MATERIALS AVAILABILITY

Parts are reserved for jobs on a priority basis. The planner enters the equipment and work order number into the terminal. Parts stocked for the equipment are displayed on the CRT screen. The planner selects and reserves the quantities of parts needed for the job. This automatically reduces available balances of part inventories. For example, with one inventory item, there are 10 units on hand. Eight have been reserved for future jobs. Two remain available for new jobs. This is all done on the screen. There is no paperwork; parts are not moved in stores. When the planned job is scheduled to be done, the planner transmits information to the stores department to pick the required parts and have them ready for pickup or delivery.

The above procedure can be followed for planned shutdown work, scheduled preventive maintenance, and routine repairs. Equipment breakdowns and urgent work require immediate availability of parts. The planner can check availability on the screen as can maintenance supervisors, craftsmen, and stores personnel. The parts must still be allocated to specific jobs. If not done this way, the available balances will be wrong. Jobs must be delayed that require greater quantities than available balances provide. When the required quantity is entered, the inventory item will show a negative available balance. The item should be ordered by stores. This will bring the available balance (on-hand plus on-order) to a positive position. When the on-order quantity is received, it will be added to the on-hand quantity. The planner can then schedule the job.

In practice, on-order items have requested delivery dates. If the dates are reliable, the job can be scheduled to begin shortly after requested delivery dates. The dates are kept current by the purchasing department and are on the screen with other parts-on-order information. For items not stocked in stores, but required for doing a job, the planner works directly with the purchasing department. Complete specifications of required items are entered into the computer system and transmitted to purchasing. Purchasing then can pull up on the screen the daily on-line requisitions for review, analysis, and conversion to purchase orders.

CONCLUSION

It would be wonderful if we all had the ideal, well-managed, on-line, integrated materials system. Most of us do not and have a long way to go. However, it is important to realize that the best investment we can make is in having reliable and accurate systems —whether manual or computer assisted.

Materials represent 50% of the total cost of maintenance. At this level, when we make improvements to the materials system, we will be making a major contribution to managing our maintenance costs.

SAMPLE SCREENS AND REPORTS

Figures 5.3 to 5.18 can serve as a guide in preparing individualized system specifications.

```
     ITEM NUMBER: ===============
       ITEM NAME: ====================
     DESCRIPTION: ==============================================
        LOCATION: ==========
   VENDOR NUMBER: ====
       UNIT COST: ======0.00
QUANTITY ON HAND: ========0
   REORDER POINT: ========0
```

Figure 5.3 Inventory control master file. (Courtesy of UNIK Associates.)

```
              ITEM NUMBER: ===============
                 ACTIVITY: -  (+/-)
                 QUANTITY: ========0
         TRANSACTION DATE: 00/00/00

  ACTIVITY CLASSIFICATION: =      A = NORMAL ACTIVITY
                                  C = CORRECTION
                  ACCOUNT: ==========
                   REMARK: =========================
              WO # / PO #: ==========


<F1> WHEN SELECTION COMPLETE      SELECT KEYS = Alt- G, E, L, R, W, C
```

Figure 5.4 Inventory control activity file. (Courtesy of UNIK Associates.)

```
  PURCHASE ORDER NUMBER: ===============        P.O. DATE: 00/00/00
     REQUISITION NUMBER: ===============        REQ. DATE: 00/00/00

  VENDOR CODE: ==========                    P.O. STATUS: =    O = OPEN
  VENDOR NAME: =========================                       C = CLOSED
               =========================
               =========================   TOTAL PART COST: ======0.00
               ==============                  OTHER COSTS: ======0.00
CONTACT PERSON: =========================        TOTAL COST: ======0.00


  COMPANY PART #:
     MFR. PART #:
            NAME:                                QTY ORDERED:
                                         RECEIVED TO DATE:
      UNIT PRICE:                               QTY RECEIVED:
       EXT. COST:                              DATE RECEIVED:

  Viewing part #                         Total # of parts in PO is

  <F1> WHEN SELECTION COMPLETE     SELECT KEYS = Alt- G, E, L, R, W, C
```

Figure 5.5 Purchase order file. (Courtesy of UNIK Associates.)

```
     VENDOR CODE: ==========

     VENDOR NAME: =========================

  CONTACT PERSON: =========================

  STREET ADDRESS: =========================

CITY, STATE, ZIP: =========================

           PHONE: ==============
```

Figure 5.6 Vendor file. (Courtesy of UNIK Associates.)

```
       COMPANY PART#: =============== NAME: ====================
 QTY-ON-HAND:              REORDER POINT:            QTY-ON-ORDER:
        VENDOR CODE       LEAD TIME (WEEKS)   LAST UNIT PRICE



QUANTITY TO ORDER: =========0        VENDOR NAME:
      VENDOR CODE: ==========     CONTACT PERSON:
            PRICE: ======0.00     STREET ADDRESS:
                                CITY, STATE, ZIP:
                                           PHONE:

     MANUFACTURER'S PART#: ==============================

                    (PRESS [F5] FOR NEXT VENDOR)

   (F1) WHEN SELECTION COMPLETE      SELECT KEYS = Alt- G, E, L, R, W, C
```

Figure 5.7 Parts to be ordered file. (Courtesy of UNIK Associates.)

```
                 MASTER INVENTORY REPORT         04/24/85      PAGE 1
COMPANY PART# LESS THAN: A


  COMPANY PART#: 253                          QTY ON HAND: 10
      PART NAME: OIL                         QTY ON ORDER: 0
       LOCATION: BIN 7                      REORDER POINT: 20
    STOCK CLASS: OH                         MAX INVENTORY: 50
    DISPOSE CAT: DS                                   EOQ: 12
     TOTAL COST: 159.50                             UNITS:
   AVERAGE COST: 15.95                   LAST PHYS COUNT: 00/00/00

VENDOR CODE     MANUFACTURERS PART NUMBER      LEAD TIME  LAST PRICE
 UK            UK-1212734/SDRI 007                  1          15.95
                                                    0           0.00
                                                    0           0.00

DESCRIPTION:
oil for lathes
--------------------------------------------------------------------
  COMPANY PART#: 4157                         QTY ON HAND: 12
      PART NAME: ASTON                       QTY ON ORDER: 0
       LOCATION: TR                         REORDER POINT: 20
    STOCK CLASS: OH                         MAX INVENTORY: 30
    DISPOSE CAT: RB                                   EOQ: 15
     TOTAL COST: 666.00                             UNITS:
   AVERAGE COST: 55.50                   LAST PHYS COUNT: 00/00/00

VENDOR CODE     MANUFACTURERS PART NUMBER      LEAD TIME  LAST PRICE
 DPLR          FPF034/340DFDF.DFOOP                 1          60.24
 UK            KSDFSDF9345,FV9045/4545              3          49.95
                                                    0           0.00

--------------------------------------------------------------------
      END OF REPORT
```

Figure 5.8 Master inventory report. (Courtesy of UNIK Associates.)

INVENTORY LOCATION REPORT 08/19/85 PAGE 1

PART NUMBER	LOCATION	PART NAME
0148624-00	05S1	8-32X2-1/4 SCREW
015167	R3S4	GASKET C700
0155662-00	05S1	BLOCK KIT
0157448-00	05S1	3/8 STREET ELBOW 90
0159182-00	05S1	RETAINING RING
0161038-00	05S1	WASHER
0162471-00	05S1	WOODRUFF KEY
019715	R2S3	PLUNGER RCT06-D2-10
01CW	P5S3	WATER HOSE WASHERS
02-0005	I4S1	VALVE BLOCK
02-0235	I4S1	RAIL BLOCK
0213390-00	05S1	BEARING NUT
0220610-00	05S1	203FF BEARING
0230236-00	05S1	RP 56 623-056 ROLPIN
0230243-00	05S2	RP-26 ROLL PIN
024275	H3S5	FILTER BAG
024287-002	F4S4	HEATING ELEMENT 440V
024287-002 240V	F4S4	240V HEATER ELEMENT
0243473-00	05S2	YOKE END

Figure 5.9 Inventory location report. (Courtesy of UNIK Associates.)

INVENTORY ON HAND 08/19/85 PAGE 1

PART NUMBER	PART DESCRIPTION	QUANTITY	UNIT COST	EXT. COST
03-0195	BODY	1	42.80	42.80
0312976-00	OIL SEAL	3	2.69	8.07
0312977-00	NEEDLE BEARING	4	1.27	5.08
0326601-00	LOCK NUT	1	0.87	0.87
0326604-00	LOCKNUT	1	1.67	1.67
034846-000	CAM ASSY	1	590.00	590.00
038441	BALL BEARING	2	19.13	38.26
041114-000	GLAND 2"	1	0.00	0.00
042403-001	NEEDLE 10-9/16IN	29	25.00	725.00
				1411.75
				1411.75

Figure 5.10 Inventory on hand. (Courtesy of UNIK Associates.)

INVENTORY ACTIVITY REPORT 08/19/85 PAGE 1

PART NUMBER	DATE	ACTIVITY		UNIT COST	CLASS	WO # / PO #
#2 GALVANIZED	07/26/85	+	2	15.00	A	BOB
024287-002	05/29/85	-	2	359.00	C	NONE
024287-002	06/05/85	+	2	235.00	A	12407
1027-9500	07/24/85	-	1	328.00	A	NONE
14	07/20/85	+	1	0.00	C	RESTOCK
1445	01/01/80	-	4	0.41	C	NONE
15	07/20/85	+	1	28.25	C	NONE
5112189	07/22/85	-	1	1726.05	C	NONE

Figure 5.11 Inventory activity report. (Courtesy of UNIK Associates.)

PARTS HISTORY REPORT 04/24/85 PAGE 1

ALL RECORDS SELECTED

PART NUMBER	NAME	# OF TRANS	USUAGE
253	OIL	93	2745
4157	ASTON	80	1822
AN200	ANGLE IRON	87	2130
B204	BELT	76	1437
BT2034	BOLT	66	1010
BT2044	CRANK ARM	70	999
BT2050	DRIVE	29	255
BT3222	YOKE	44	760
T204	TRANSFORMER	34	398

END OF REPORT

Figure 5.12 Parts history report. (Courtesy of UNIK Associates.)

ACCOUNT HISTORY REPORT 08/19/85 PAGE 1

ACCOUNT	# OF TRANSACTIONS	AMOUNT
7500	1	30.00
7536	2	1188.00
7540	1	328.00
7542	1	1726.05
7564	3	29.89

Figure 5.13 Account history report. (Courtesy of UNIK Associates.)

```
                    VENDOR REPORT
ALL RECORDS SELECTED

VENDOR CODE         NAME                         ADDRESS
-----------         ----                         -------
DPLR          DOPLER MANUFACTURING          125456 EAST NOKOALE
JOE TERNAME                                 ST. PAUL MN   50040
                                                 123-321-1239

UK            UNIK ASSOCIATES               12545 W. BURLEIGH
KISHAN BAGADIA                              BROOKFIELD, WI 53005
                                                 (414)-782-5130

UTAD          UNITED TOOL AND DIE           12344 N. SLIPPRIE
FRANK GREASEN                               BOOMSTOWN, OHIO  43333
                                                 (555)-549-9472

       END OF REPORT
```

Figure 5.14 Vendor report. (Courtesy of UNIK Associates.)

INVENTORY BELOW REORDER POINT REPORT 08/19/85 PAGE 1

PART NUMBER	PART NAME	QTY-ON-HAND	QTY-ON-ORDER	REORDER POINT
015167	GASKET C700	1	0	2
10143	NEEDLE ASSY	98	0	100
102-792	100 MESH FILTER SCR.	0	0	2
10250T172	JUMBOHEAD SWITCH	0	0	1
2000-25	SUCTION COVER PLATE	1	0	2
2000-35	MANIFOLD GASKET	1	0	8
203B	203B DEMA VALVE	0	0	2
2065	THRUST WASHER	0	0	2
20H	2/0 WIRE SPLIT BOLT	0	0	10
226	FUSE REDUCERS	2	0	3
30225	AAE-A201 RELAY	11	0	12
320-2110	SEAL	1	0	2

Figure 5.15 Inventory below reorder report. (Courtesy of UNIK Associates.)

OPEN PURCHASE ORDER REPORT 04/24/85 PAGE 1

ALL RECORDS SELECTED

PO NUMBER	DATE	VENDOR CODE	STATUS	TOTAL COST
P-0340-03400	04/24/85	UK	0	483.48
P-2303-0005	04/24/85	DPLR	0	903.60

END OF REPORT

Figure 5.16 Open purchase order.

YOUR COMPANY NAME HERE
ADDRESS
CITY, STATE, ZIP
TELEPHONE

PO #: P-0340-03400
DATE: 04/24/85

To:
UNIK ASSOCIATES
12545 W. BURLEIGH
BROOKFIELD, WI 53005
(414)-782-5130
KISHAN BAGADIA

QUANTITY	MFR PART NUMBER	NAME	PRICE
12	UK-1212734/SDRI 007	OIL	191.40
25	LFG9DFGLDFGOFDG90RTE	BELT	225.00
12	KLMDFGIOJERFG90DFGDFG/DF/001	BOLT	67.08

Total part cost : 483.48
Other costs : 0.00
Total cost : 483.48

END OF REPORT

Figure 5.17 Typical purchase order form. (Courtesy of UNIK Associates.)

PARTS RECEIVED REPORT 08/19/85 PAGE 1

DATE RECEIVED: 08/17/85

PART NUMBER	QUANTITY	PART NAME
0148624-00	100	8-32/2 -1/4 SCREW
0155662-	3	BLOCK KIT
0159182-00	6	RETAINING RING
01CW	12	WATER HOSE WASHERS
024275	12	FILTER BAG
024287-002	3	HEATING ELEMENT 440V

Figure 5.18 Parts received report. (Courtesy of UNIK Associates.)

6

Software Selection

INTRODUCTION

This chapter deals with some basic principles of selecting maintenance software. It will help you to evaluate the requirements of your company and find the right software to meet your needs. It will help you choose programs that are flexible enough to conform to your way of doing business—programs that should ultimately help your business become more productive.

In choosing maintenance software, the most important factor is knowing your own business—your own requirements—so that you know what to look for. This chapter should help you make your business more productive by choosing the right maintenance solution. That can be done only if you take the time to

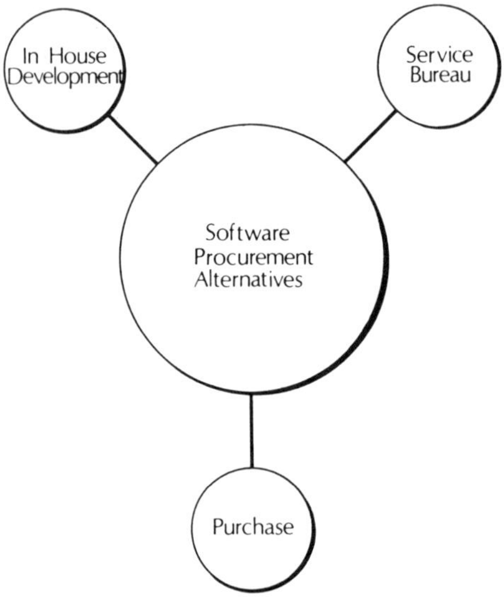

Figure 6.1 Software procurement alternatives.

study your requirements and select the software that best meets your needs.

Once a computerized maintenance management system has been budgeted, the next step is to select a maintenance software package. Where do you start and how do you proceed?

Basically, there are three alternatives (Figure 6.1): (a) in-house development, (b) service bureau, and (c) purchase software.

IN-HOUSE DEVELOPMENT

Before the relatively recent explosion of packaged software onto the market, some companies found it necessary to develop their own maintenance software packages. Generally speaking, these companies were one of two types: a large multiplant operation with a great many resources, and/or an industry so specialized as to have difficulty finding the right software package for its needs.

Ideally, a customized maintenance software package or in-house system offers a great deal of flexibility to a company. It also provides the best link with existing plant information systems, because it can be designed to accommodate the needs of other departmental systems. And the in-house system can be designed to meet the highly specialized needs of a company, while readily adapting to its current maintenance operations.

Today, however, relatively few organizations are opting to develop a maintenance software package of their own. There are at least two good reasons why they do not go this route. First is the high cost, and second the long development time. Those companies that have developed their own system probably spent several years and hundreds of thousands of dollars in its making. No one should underestimate the magnitude of such a task.

It is important that maintenance information be entered, queried, and displayed on-line, but it is equally important that maintenance personnel and foremen have the information they need on a form or piece of paper they can carry with them to the job site, and on which they can make notes. This paper may serve as a turnaround document, or it may lead to the issue of more paper and later on-line entries, after discussion with the foreman. It is unrealistic to expect a maintenance person or foreman to carry a CRT terminal when inspecting or repairing equipment in difficult-to-reach areas. An ideal, purely paperless on-line system, where everything is handled on CRT screens, is not a practical solution to maintenance management.

Summary

Major advantages of in-house developed system:

Will provide the greatest amount of flexibility in program development.

Will best link with existing plant information systems because it can be designed to accommodate the needs of other departmental systems. This flexibility is likely to improve acceptance of the system by managers.

Will meet highly specialized needs.

Will be readily adaptable to current maintenance operations.

Major disadvantages of in-house developed system:

High cost.

Long development time. Generally, the computer programmer is unfamiliar with maintenance operations and must develop an understanding of the maintenance environment. A maintenance manager may be unfamiliar with computer capabilities and programming. Thus, the learning process often requires design-redesign cycles and several program changes before a system is ready for use.

SERVICE BUREAU

Another alternative open to a company looking for solutions in the area of maintenance management is the service bureau. The term "service bureau," in general, refers to an outside agency, either local or mail order, that monitors a company's equipment and maintenance schedules. It provides all equipment and personnel necessary to produce the required reports. The user has only to provide the necessary input data to the bureau and wait for the results to be returned. The service bureau provides a researcher to search maintenance equipment files and collect the needed preventive maintenance (PM) and other related maintenance data on each piece of equipment to be included in the program. This service is important to a plant engineering organization unable to devote employee time to accomplishing this task. Then, in turn, the service bureau generates work orders and produces the feedback or periodic reports needed by the maintenance staff.

After the PM and other maintenance tasks are completed, the work orders are reviewed by the maintenance supervisor and returned to the service bureau. The updated information from the completed work orders is entered into the computer, and periodic reports are generated for individual pieces of equipment. With these reports, steps can be taken to prevent unforeseen equipment failures, and important management decisions can be made. It is important to note that the effectiveness of these reports depends largely on the written comments of the craftsmen. For the program to be successful, each individual craftsman needs to see him or herself as an integral part of the system.

Unlike the in-house system, this method of managing maintenance functions requires little or no training or investment in hardware or software; it usually costs less to install than the other two alternatives; and it takes a relatively short amount of time to implement.

However, there are some negative aspects to the service bureau. Based on a per machine rate, the enlisting of a service bureau does entail an on-going service fee to the customer. Simply put, the more machines a company has, the greater the fee will be to maintain its records. In addition, it may have difficulty linking up with existing plant systems or programs or meeting a company's specialized needs.

It might be worthwhile exploring a service bureau only to see whether the vendor would be willing to provide service for a limited time. This way a company may determine, without much cost, the right software for its operations. However, not all vendors will be willing to go this far. It should also be noted with the cost of hardware and software continuing to decline, the cost effectiveness of the service bureau is also decreasing, making it less of a bargain.

Summary

Major advantages of a service bureau:

- Low installation cost.
- No investment in computer hardware.
- No software training required.

Major disadvantages of a service bureau:

Ongoing fee based on number of pieces of equipment.

Will not link with existing plant systems.

May be difficult to meet specialized needs.

If the equipment at the bureau fails and if they do not have adequate backup equipment, it may be difficult to obtain critical reports.

PURCHASE SOFTWARE

As already indicated, both the in-house and service bureau alternatives have at least a few serious problems.

To quickly review, a customized in-house maintenance management system could take a year and cost as much as $100,000 or more to develop. Although it is true that a service bureau does not require nearly as much time to set up as an in-house system, it may have difficulty meeting specialized needs of a company or linking up with existing systems and may be costly over the long term. With these points in mind, it's time to look at a third alternative—purchasing software.

The purchasing of software, in this case, simply refers to buying a ready-made maintenance software package from a firm specializing in the development of such packages. Research has shown that the purchasing of software from a vendor for a maintenance management system can offer significant benefits to a company, especially in the savings of time and money. In most cases, the software package can be installed, tested, and put into operation by a user in less than a month. Since the software package has been developed by the vendor as a solution to the problems of several companies, it may incorporate more sophistication than even the company's in-house system.

Summary

Major Advantages of Purchased Software

Relatively low cost.

No development time.

No ongoing fee based on number of pieces of equipment (as in service bureau).

Can be implemented in a short period of time.

A good software package will provide state-of-the-art technology.

Will probably provide more features than your proposed system, thus further improving productivity.

A well-developed software package will provide the required flexibility.

Major Disadvantages of Purchased Software

May not link with existing plant information systems.

May not meet highly specialized needs.

(See Table 6.1 for advantages and disadvantages of a packaged software.)

This alternative brings up the question "How do I find and select the maintenance software that is best for my application?"

If the proper selection method is not followed, it may result in acquiring software that may not satisfy the company's needs, delayed implementation, canceling projects, returning software, and frustration.

There are probably a number of ways to select a software package. You could rely solely on those programs that you see advertised in popular trade journals. You can call a competitor of yours and ask him about the system he bought. You can hire a consultant to figure out what you would need, based on your requirements. No matter which way you choose, there are some criteria that you should keep in mind.

Table 6.1 The Pros and Cons of Applications Packages

Positive factors	Negative factors
Less programmer time	Suitable package not available or can't be located
Less duplication of effort	
Faster implementation	Possible inefficiency
Superior product	Need for tailoring
Lower cost	Inflexibility
Lower maintenance cost	High capital outlay
Better support	Lack of support
Better documentation	Poor documentation
Staff freed for other work	Lack of acceptance in the installation

Note: Items appearing in both columns illustrate the fact that they are relative and can vary from one package to another and from one installation to another. These points are discussed in detail in the text.
Source: S. Wooldridge, *Software Selection*, Auerbach Publishers, Inc., Pennsauken, N.J., 1973.

DETERMINE THE OBJECTIVES

The first step is to develop a complete outline of the scope, goals, and objectives of the proposed system.

Follow each of the following steps to assure adequate attention is given to the first and most important phase of the project:

1. Identify the major problems to be solved by the implementation of computerized systems. As a double-check on the real priority of these items, roughly quantify the benefits these improvements will yield. How much money can be saved by achieving the selected objectives?

Defining current system problems will provide management with evidence of the need to resolve them. It is important that problems be well documented in terms of lost revenue, inefficiency of operation, and inability to meet current and future business goals.

Management chooses maintenance software for many reasons: to save money, to keep ahead of the competition, to improve productivity, to increase efficiency, to expand their businesses, or to provide quick access to maintenance information for managerial decision making and planning. Be aware of which of these benefits are important to you.

The major benefits will vary from plant to plant, industry to industry, and organization to organization. In one situation, downtime on production equipment may be the primary concern, whereas, in another, it may be the productivity of the maintenance personnel. The system should be specified to achieve the maximum economic advantages in an individual situation.

2. In the development of the needs and objectives for the new system, input should be obtained from every organization and level that will interact with it. All levels of maintenance management should have an opportunity to make their needs known and make their contributions. This is especially true for first-line maintenance supervisors.

Unless the system being scoped is strictly internal to a department, there should be input from other departments. Production personnel should be consulted about proposed maintenance systems since they are almost always the primary customers.

Other departments that may have an interest and be able to make contributions to the planning include purchasing, receiving, cost accounting, payroll, and manufacturing engineering.

3. It is important that the system being considered is not limited to a computerized version of existing procedures because that might prevent exploring additional features which would further improve maintenance productivity.

To ensure that new ideas are being fed into the planning process, the following steps are recommended:

1. Thoroughly review current periodicals for new ideas that are working in other similar plants and organizations.

2. Review the features offered by the largest and most sophisticated computer software application package for good ideas—even if the packages are far too sophisticated and expensive for your application.
3. Attend seminars covering the latest developments in maintenance management.
4. Discuss your situation with reputable consultants and other experts in the field of modern maintenance management.
5. Contact noncompeting firms in other industries that have advanced systems operating in environments similar to your own.

If followed thoroughly, these rules or suggestions should provide the basis for your maintenance improvement plans for the next 3-5 years.

IDENTIFY THE HARDWARE ALTERNATIVES

The ideal situation is to match the hardware to the most suitable application package available. Normally, this degree of freedom is not available to the maintenance department. Company policy or existing computer facilities may dictate that all programs shall reside in the central computer. In other cases, decentralized data processing may be acceptable but the type and make of stand-alone hardware may have been established.

It is strongly recommended that you work as closely as possible with the MIS (maintenance information system) group. Listen carefully to their recommendations especially with regard to technical matters such as memory capacities, processing speeds, compatibilities of interfaces, etc. MIS, like maintenance, is a service organization trying hard to keep their customers happy in a rapidly changing environment.

DEVELOP THE SYSTEM SPECIFICATIONS

With objectives and a general outline of the system agreed upon, the next step is to define, in detail, the specific features desired of the new system.

1. Identify the major aspects of the system.
2. Clearly define all significant features.
3. Group the features within each subdivision into three categories of importance:

   mandatory — features of such importance that a package in which they are not fully provided would be unacceptable.

   needed — features that are of real value but the omission of a few scattered items does not indicate outright rejection. Their omission does reduce the value of the system.

   desired — features that are of only specialized usefulness. Or they may be of marginal value. These are features you would like to have but can do without.

Values then can be assigned to each feature within these latter two categories. A scale of 1-10 is suggested with needed features ranging from 4 to 10 and desired features from 1 to 3. See Appendix A for system specifications examples.

SEARCH FOR APPLICABLE VENDOR PACKAGES

In this very specialized field of maintenance, locating appropriate, practical, and proven systems for the various aspects of control is not easy. The standard sources, such as general software directories, are of little help. (See Appendix A for system specifications examples.)

Professional societies like the AIPE can be of real assistance. Reading both ads and articles in professional journals will provide other leads. The very best source will be to make best use of your time at seminars and conferences. If your time is limited, gather all the detailed literature you can find. Then, as time permits, review and discuss with vendors the details of those systems which, on first look, appear to be best suited to your needs.

CONDUCT PRELIMINARY SCREENING FOR THE TOP CANDIDATES

When the initial system's literature has been gathered, a quick preliminary review should be made as follows:

1. List the system name and vendor, compatible hardware, and language.
2. Quickly review the available literature to determine how many of the mandatory features appear to be provided by each of the candidate systems.
3. Eliminate those systems which are not compatible with the available hardware or language and don't meet majority of the mandatory features.

Review the top three or four candidates with the following criteria in mind (see Figure 6.2).

Flexibility

The software should be flexible in terms of allowing you to enter information pertaining to your organization. A few examples are cited here:

1. It should permit you to enter your own coding for craft personnel and provide the capability to change in the

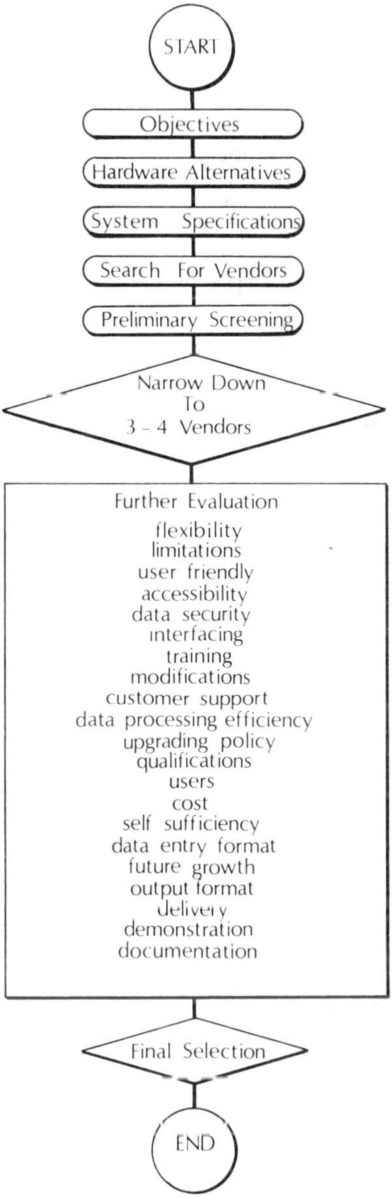

Figure 6.2 Step-by-step procedure for software selection.

future as opposed to locking into their own coding system. For example, the software might ask you to enter 01 for mechanical, 02 for electrical craft, etc. This is locking you into a system. If you have been using ME for mechanical, it should allow you to enter that so you do not have to change your existing system.

2. Priority in a *work order system.* It should allow you to enter your own coding for priorities.

Limitations

Check the limitations, if any, on the number of records you can store. For example, you may have 10,000 parts in your inventory and the software may limit you to entry of only 8000 parts. Always allow for future expansion as well. If you have 10,000 parts at present, determine the number you would have a few years from now. You do not want to invest a great deal of time and effort in software that will not work after a short period of time.

(If the software presents limitations, ask the vendor if it can be modified to overcome them.)

User-Friendly Software

The key to successfully computerizing your maintenance systems is to select user-friendly software. If the software is difficult to use, you may have a hard time getting it to do what you want it to do. But if your software is user friendly, the maintenance applications will be easier to use. Your employees will not be frightened by the system and should enjoy learning to use it. And you will probably have far fewer headaches making the change from paper to computerized system. Constant reference to the user's manual to successfully execute the program becomes excessively time-consuming and leads to frustration and mistakes.

A software package must have the following features in order to be truly user friendly:

Easy to Learn Programs

Some programs are easy to use once you know how, but learning them can be an agonizing experience. The problem is that a bad learning experience can cause your employees to become frightened of the computer. And if they are "turned off" by the system, they won't be enthusiastic about using it.

Another advantage of easy-to-learn software is that no one computer operator becomes an indispensable employee. If the operator has to leave the job, you can more easily train a new employee.

Here are some of the ways in which software manufacturers provide training to people who use their systems: Software tutorials are self-teaching instruction programs you complete at your own pace, using your own computer system. These tutorials can include workbooks, audio cassettes, and special "learning diskettes" that display training instructions on the computer screen.

In addition to presenting instructions on the screen, sophisticated tutorials also quiz you as you learn. The system will require you to review areas you have trouble with and will retest you on them until you are familiar with the concepts. This technique is known as computer-assisted learning or programmed instruction.

The tutorials also include practice problems. These problems should take you from a basic understanding of the system to detailed instruction in the system's various features and options. Be wary of tutorials in which practice problems cover only the basics and leave you to figure out the advanced concepts on your own. It is these advanced concepts that you probably need the most help with.

Software training materials should be clearly written and well illustrated. Drawings, charts, and pictures aid in learning and serve as quick references for refreshing your understanding of concepts and procedures. The instruction materials should be written in plain, simple English with a minimum of jargon.

Some software vendors supply training material in textbook or workbook form without specific practice problems or self-test questions. If so, you should develop your own practice problems

UNIK ASSOCIATES
MAINTENANCE MANAGEMENT SYSTEM
MAIN MENU

A EQUIPMENT DATA

B PREVENTIVE MAINTENANCE SCHEDULING

C WORK ORDER SYSTEM

D LABOR CRAFT DEFINITION

E INVENTORY CONTROL

Q QUIT PROGRAM

Figure 6.3 Main menu. (Courtesy of UNIK Associates.)

for training your employees. Keep the problems simple, but make sure they cover all the functions.

Menu-Driven

A menu is a list of program functions displayed on the computer's screen. Figure 6.3 shows the menu for a maintenance management software package.

Each function in the menu is labeled with a number or letter. To select a function, you press its number or letter on your keyboard. The computer then puts that program function to work.

Menus provide a quick and easy way of telling the computer what you want it to do. In programs without menus, you may have to enter the name of the individual program function and several complicated commands in order to get the computer to perform that task. In simple applications, a single menu causes the computer to run each job or series of programs in a preplanned

EQUIPMENT DATA MENU

A VIEW/UPDATE AN EQUIPMENT RECORD

B ADD AN EQUIPMENT RECORD

C CURRENT USAGE UPDATE

D DUPLICATE AN EQUIPMENT RECORD

E DELETE AN EQUIPMENT RECORD

F PRINT EQUIPMENT DATA REPORTS

@ RETURN TO PREVIOUS MENU

Figure 6.4 Equipment data menu. (Courtesy of UNIK Associates.)

order. But in maintenance software and other sophisticated applications, the computer needs more than one menu to describe all the jobs the software can do.

In these applications, a main menu lets you choose one of several secondary menus. Figures 6.4 through 6.7 show the secondary menus for functions A, B, C, D, and E of the maintenance application displayed in Figure 6.3.

If the application can perform a large number of tasks, selecting a function on a secondary menu may lead you to an even more detailed submenu. Menus usually contain commands that let you get back to the main menu, in case you want to start over or choose another function. A logical organization of main menus and secondary menus makes it easy to do your maintenance job quickly without having to refer to a manual.

PREVENTIVE MAINTENANCE MENU

A AUTO-GENERATE WORK ORDERS

B LOG COMPLETED PREVENTIVE MAINTENANCE

C PREVENTIVE MAINTENANCE REPORTS

D ACCESS PM MASTER FILE

0 RETURN TO PREVIOUS MENU

Figure 6.5 Preventive maintenance menu. (Courtesy of UNIK Associates.)

Input Screens

When you want to enter information into the computerized maintenance system, the computer displays an input screen. The input screen is a blank "form" displayed on the display screen (monitor). You use the keyboard to fill in the blanks and enter the appropriate information. (Figures 6.8 and 6.9 show sample input screens for equipment data and preventive maintenance files, respectively.)

Input screens are designed to make it easy for operators to enter information correctly. Well-designed input screens help to simplify data entry and reduce operator errors. Here are the features to look for in a program's input screen:

Uncluttered screens.

Information entered into the system is placed in the same position on both the screen and the paper forms you use for manual collection of the data.

WORK ORDER MENU

A VIEW/UPDATE/COMPLETE AN EXISTING WORK ORDER
B INITIATE A NEW WORK ORDER
C DUPLICATE A WORK ORDER
D CANCEL A WORK ORDER
E PRINT WORK ORDERS
F WORK ORDER REPORTS
Q RETURN TO PREVIOUS MENU

Figure 6.6 Work order menu. (Courtesy of UNIK Associates.)

The screen provides the user with easy-to-understand instructions for entering information.

The system allows the operator to review and correct typed information before it is stored in the computer's files.

For some sections of program input, default values are desirable to speed program execution. For example, if date is a required input, today's date can be entered by the user with a carriage return.

Error Handling

Computer users make mistakes. User-friendly software can tell when certain types of errors have been made and then display warnings or error messages on the screen.

Error messages should be worded clearly and directly so that the operator is sure to understand their meaning. Also, the user

```
INVENTORY CONTROL MENU

A   ACCESS INVENTORY MASTER FILE
B   ACCESS INVENTORY ACTIVITY FILE
C   PRINT INVENTORY REPORTS
0   RETURN TO PREVIOUS MENU
```

Figure 6.7 Inventory control menu. (Courtesy of UNIK Associates.)

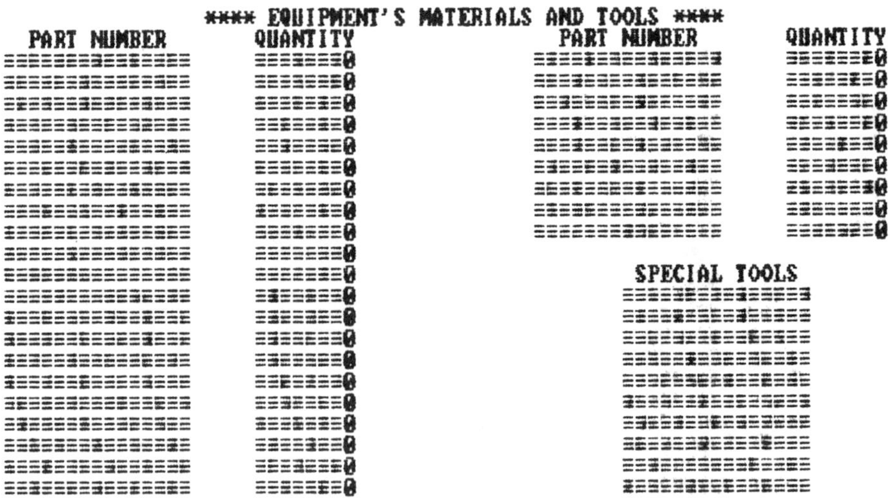

Figure 6.8 Equipment data input. (Courtesy of UNIK Associates.)

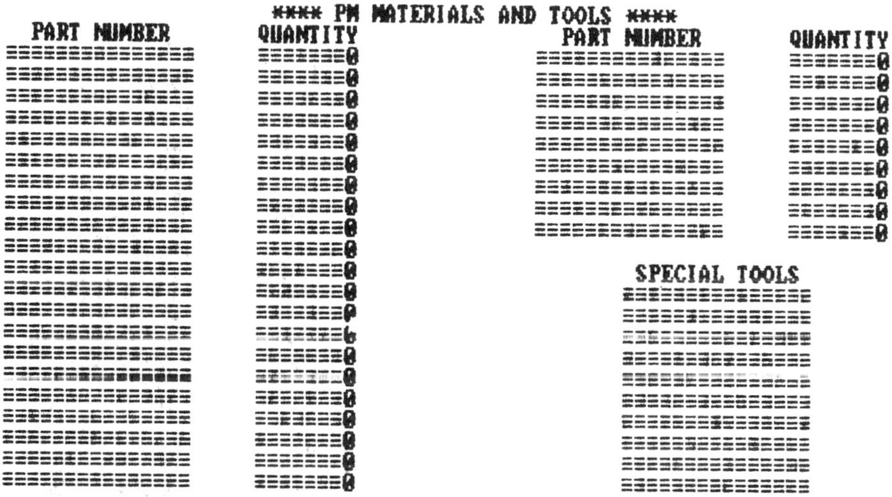

Figure 6.9 Preventive maintenance input. (Courtesy of UNIK Associates.)

manual should contain a list of error messages and their explanations.

In addition to defective errors, user-friendly software guides operators through the process of correcting their mistakes. Errors can be corrected quickly and easily, without the need to redo a lot of work or follow complex procedures.

Help Text

Computer users sometimes get stuck when working with an application. But finding the answer in a manual or getting help from an expert takes time. That is why user-friendly software provides help text. Here's how it works:

Let us say you have a question on the function you are working with. By pressing certain keys on the keyboard, you can command the computer to display instructional information on that function. This information can help to answer your questions, thereby allowing you to continue with your task.

Printer Interrupt

User-friendly software lets you stop the printer and then restart it where you left off—without having to reprint the entire report or do a lot of extra work. This feature is helpful if you need to stop the system in the middle of a long report or have a problem with the printer.

Accessibility

Find out whether the vendor can provide some means of limiting the access to selected personnel such as a "password system." A multiple-level password is recommended. For example: One level could be top management, which has access to all the files. Another level could be keypunch operators, who may add or inspect a record, but not update it.

Data Security

The program should provide some backup facility to protect against loss of data through accident or otherwise.

Interfacing

The program should have some provision to share appropriate data with another computer, for example, provide a machine-readable record of parts cost to the accounting computer.

Training

Find out whether the vendor has a provision for training, either at his facility or on site. It is always a good idea to go through a training session provided by the vendor. In the long run, the small investment will save a great deal of money and frustration.

Apart from user training for data entry clerks and planners, it is desirable that training seminars be available at specific locations, or at the user's plant, for all maintenance supervisors and personnel, prior to startup. This is to ensure that the system will really work well on the shop floor, rather than just in the data-processing department.

Investigate the following:

Training cost

Number of training days

Training location

Cost of additional training, if required

Modifications

Each company operates a little bit differently. A very sophisticated feature in the software for one company may be a limitation for another. The vendor should be able to make the custom modifications to suit your application. If you are going to invest the money, it might as well be for something that you really want and is going to work for you. (It becomes more significant as the majority of the vendors don't supply the source code.) In many instances, using the software for 3-6 months provides you with the opportunity to go into fine details of the program. Now you are getting smarter and are capable of thinking about possible enhancements in the program that will further improve your productivity. At this point, you may be approaching the vendor to explore the feasibility of making those modifications. That is why it is important for you to make sure the modification service is available.

From vendors who do not provide a copy of the source code, a written contract should be obtained stating that a copy of the source code should be kept with an attorney or some other escrow agent, and the copy of the source code will be made available to the user in the event something happens to the vendor (going out of business, filing bankruptcy, etc.). See Appendix B for a typical contract.

Customer Support

Investigate whether the company has proper technical support and make sure there is someone available to answer your questions. This is important, as in the beginning you may require very extensive customer support.

This support should include: (a) a telephone response line you can call for answers to your questions; this service is usually available for a fee when the warranty runs out; (b) newsletters that provide tips and advice on how to get the most out of the application package.

Data-Processing Efficiency

You will be storing thousands of maintenance records. The response time is very critical. Check to see which programming language has been used by the vendor. A program written in a high-level language (C, assembly, or PASCAL) is significantly faster than one written in basic.

Upgrading Policy

Does the vendor have an upgrading policy? If so, what kind? A good software company periodically upgrades its products based on feedback received from users and potential users. Generally, the enhanced version costs more than the original. Investigate the company's policy. An acceptable policy is to charge the difference. For example, if the old version was $1995.00 and the upgraded version is $2495.00, you should pay the difference of $500.00. These upgrades can be corrections of problems or improvements that enhance the system's capabilities. The vendor should also send documentation that explains these changes and improvements.

Qualifications

Check the qualifications of the vendor staff. What kind of experience and background do they have pertaining to maintenance management and computer software and hardware. Remember, you want to deal with people who are knowledgeable in the field, who can offer you the help and assistance you need to improve your maintenance operation.

User List

Telephone interviews should be conducted with a number of users of each vendor's system to describe their experience and satisfaction with the package.

1. Prepare an outline of the information to be gathered and the specific questions to be asked of each user.
2. Obtain a comprehensive list of users from each vendor—preferably a list of all users. Do not get only a selected list of a few of the vendor's favorites.
3. In making your selection of users to contact, follow the vendor's suggestions for two or three, and then pick several more on your own. Perhaps these would be from your industry or companies of your size. If in the local area, there may be an opportunity to visit the user, if appropriate.
4. Try to talk to at least two different people at each location to minimize individual biases. It is desirable to talk with both a system user, a person in maintenance management, and a person in MIS who supports the system.

There may be some value in visiting other installations where the maintenance software package under consideration is already being used. The visits, if they are made, should be considered as a secondary source of information and background. It is usually best that the supplier make this initial approach, and try to choose an installation with a similar operation.

Cost

This is a major criterion. You should investigate all possible cost factors associated with the software, such as:

a. Software acquisition cost
b. Implementation cost (e.g., installation)
c. Training cost
d. Maintenance cost (ongoing)

Make sure there are no hidden costs involved.

Self-Sufficiency

Programs should be capable of direct, full use without or with very little need to consult a manual or other outside sources. Instructions on the screen should explain what the program will do and how to use it.

Data Entry Format

Programs should be largely menu-driven so options can be selected by a single keystroke. Commonly used data should be stored in permanent files or provided as default values. Data entry routines should allow correction or editing of data after initial entry without restarting the routine. The program should be capable of checking data consistency and trapping errors. Good computer programs have built-in checks on data, which will reject invalid entries. A program cannot find every error but can determine whether values are within an allowable range; that is, a day of the month should fall between 1 and 31.

Future Growth

By deciding on software that can meet your growing needs, you help to ensure that your maintenance package will not be obsolete in a year or two. The software you select should be capable of handling both your current volume and your projected volume for some time to come. Have these figures handy when you shop around; different maintenance packages have different limits.

If you do not need to automate all maintenance functions now but think you might automate more in the future, choose maintenance software from a family of applications that can be installed one at a time. In these packages, each application can run independently of the others and can be integrated when desired. This way, you can start small, then automate additional areas as the needs arise. This way you avoid acquiring software that you have no immediate use for. For instance, you might start by automating preventive maintenance. When you add the work order system later, it can use information directly transferred from the PM package.

Output Format

The program should provide results in final form without need for further processing. Repeat format should be flexible, allowing choice of summaries or detailed reports.

Delivery

Is the package available now? Do not depend on a package that is supposed to be available on a certain date. Computer program expected-completion dates have been known to be very inaccurate, even those of giant companies such as IBM.

Demonstration Package

Check to see whether the vendor has a provision for demonstrating the software. You should get a chance to review the software before purchasing. The idea is to make sure the program performs as outlined in the sales literature.

Your investment is not just in the software. You will be investing a significant amount of money in hardware, training, and implementation of the system. Therefore, it is important to ensure you are investing in the right package.

The more the demonstration is like a live run, the more valuable it will be. Remember, however, that the more control the supplier has over the demonstration—if he sets it up and has his own poeple operate it—the easier it is for him to show the best features of the software and gloss over the worst ones. Use live data if possible, or if not, prepare test data in the installation.

You should also ask how long you have to evaluate whether or not the package actually works in your business. Remember it takes time to learn to use a program. If you are strapped into evaluating a package in just a few days, you may find yourself rushing. In practice, it can take a number of weeks for proper evaluation.

Documentation

As microcomputers become more popular, software is reaching less-experienced buyers. More than 40% of new buyers have little or no experience in using microcomputers. Software user manuals are the most important link between a microcomputer and a novice user. A well-designed software manual can help a new user overcome technophobia and enter the world of technology.

The software should be accompanied by an instruction manual. As a matter of fact, it is highly recommended to review the instruction manual also before buying the software. This is the most important piece of documentation.

COMPILE RESULTS, COMPARE SOFTWARE PACKAGES, AND SELECT VENDOR

With all steps of this evaluation procedure completed, you are ready to compile the results and make comparisons.

Another point to remember, a price guarantee clause, stating that the vendor must guarantee its price for 120 days after the date the proposal is submitted, should be obtained.

1. Compile the results from each of the four areas: (a) system features, (b) vendor profiles, (c) user experience, (d) cost. The vendors and packages should be ranked in each area: first, second, third, and so on.

2. Compare these final rankings. A direct mathematical comparison is not recommended because the four categories dif-

1. Reputation and business integrity
   Business areas
   Age of company
   Clients
2. Size of the company
   Number of employees
   Number of sites
   Expansion plans
3. Financial soundness
   Capitalization
   Profitability
4. Methods of working
   Standards used
   Project-control techniques used
   Staff assignments
5. Expertise
   Background of managers
   Technical ability of systems and programming staff

Figure 6.10 Summary checklist for evaluating the supplier. (From S. Wooldridge, *Software Selection*, Auerbach Publishers, Inc., Pennsauken, N.J., 1973.)

fer so much in importance, character, and precision of measurement. In most situations, the relative importance given to the various areas is as follows: (a) system features, (b) user satisfaction ("Ask the man who owns one"), (c) total system cost, (d) vendor profile (see Figure 6.10).

A large measure of good judgment is still necessary despite all efforts at quantitative comparison. However, the judgment will be based on a sure footing of factual data.

3. Select the vendor who provides the best combination of characteristics for your particular situation. If two or more systems are approximately equal, so much the better. Now enter into serious discussions with the vendors and attempt to eliminate the weakest characteristics of each of these systems through negotiation. It may be good to ask for assistance from your purchasing department and MIS. Often they can play the "tough guy" role by demanding price or technical concessions while you, as the user, maintain a smooth relationship with the potential vendor.

Beware of selecting a software package purely on the basis of price or data-processing satisfaction. The software price and computer running cost will usually be far less than your investment in training, implementation, and continued operation on the shop floor.

Top management's evaluation of a computerized maintenance management system is based on whether the maintenance personnel are convinced it helps them, it works well on the shop floor, and it actually cuts overtime and contract labor and reduces backlog and downtime.

CONCLUSION

The primary objective of this chapter has been to share with you the importance of the software decision. We have discussed topics such as:

How does a maintenance software program work?

What is important about the selection process?

Why should you involve your employees?

What features does a quality maintenance software package provide?

The well-conceived and executed introduction of computer-based systems into the management of the maintenance function can bring about major advancements. Improvement will occur in both the efficiency and effectiveness of the operation.

The conception or planning stage is paramount to the success of the overall program. The most widely selected packaged system will be of modest benefit if the planning phase has not correctly identified the key needs of the department. Make sure that the planning is thoughtfully and creatively done. Identify the real factors providing long-range leverage to costs. The result should be a modern computer-based maintenance management system that will quickly pay for itself, for years to come, through savings in costs and improvements in the service provided.

Your level of satisfaction with the software that you choose may largely depend on your involvement during the selection and installation process. Your continued involvement should help you achieve results to better meet your needs.

7

Justification

INTRODUCTION

Most managers find it increasingly difficult to control rising maintenance costs because of inadequate or outdated procedures. It is now possible for managers to purchase a microcomputer-based maintenance system that can provide all the needed information about a plant's maintenance operations. The low cost of microcomputers puts them within reach of many small maintenance shops. Recently, microcomputers have become so economically attractive that no maintenance function can ignore their advantages. Paramount among the advantages is the fact that the computer belongs to maintenance and can be used in real-time mode. These microcomputers can be placed on a regular desk, minimizing the space requirements, and they do not need any special environmental conditions. However, before considering the purchase of such a system, the company has to justify it, which basically means convincing people. How do you convince different levels of management? Simple—you have to learn to speak their language.

Management will be receptive to any well-planned cost reduction program, so the best approach for a maintenance program justification to management is one that emphasizes dollar savings. And if a higher-quality product can be produced for less money, everyone connected with the facility will benefit. Usually, the job of selling such a program must be aimed at two distinct levels of plant operating personnel: department managers and line supervisors.

Probably the most difficult part of starting a computerized maintenance program is convincing management that savings will result. Maintenance programs may not show immediate savings. In fact, in almost all cases, costs will show slight increases, which can be attributed to the expense of starting a new program while the old emergency repair system is still in effect. It should be clarified here that a well-planned and -run manual system will also result in improvements. The activities of planning, diagnostics, optimum order quantities, inventory control, preventive maintenance scheduling, and so on, are not limited to computerized systems. By using the computer, further benefits can result.

Because of this initial negative appearance of maintenance programs, it is necessary to show the long-run benefits. The plant engineer should show: (a) the actual costs for the 2 years immediately preceding introduction of the maintenance program, and (b) projected costs for the 2 years following startup of the program. This kind of analysis will show what the program can do, and it will provide the plant engineer with a budget reduction goal. Program projects must be realistic, however. If the figures are too extreme, management may simply elect to kill the program before it even gets started. The plant engineer must plan thoroughly so that he is sure of his position before he approaches management with his idea.

Long-range planning is not the most significant factor when discussing the maintenance plan with first-line supervisors and production management. In most cases, because production is asked to produce more than the facility is capable of producing, the production manager must constantly seek ways to improve the operation. Even though equipment failure adds to his already overburdened workload, he is often reluctant to listen to any suggestions to stop an operation simply for something like preventive maintenance. Therefore, the plant engineer must convince the production manager that the program is designed for his benefit.

One last group must be convinced—the maintenance men themselves. Maintenance mechanics frequently resist such a program, because they believe that equipment should not be touched as long as it is still running.

The best approach for convincing maintenance personnel is the proof-by-results method, which may be described as a visible and unquestionable happening, where the end observations show evidence of time and money savings.

MISCONCEPTIONS

Following are some facts that every plant engineer should know before attempting to justify a computerized maintenance management program. Before a computer is used, the plant engineer must be careful to emphasize that the computer can't solve all maintenance information problems and that an extended debugging and testing period may be required once the program has been implemented. The management must be made aware of the extended period before the program begins to carry its own weight.

Such a program will not reduce maintenance manpower at least not until the program has been in operation for a period of time and working smoothly (amount of time needed for implementing the program will vary with each application). In fact, an effective maintenance program may create an immediate need for an increase in the number of maintenance people. The belief that an effective maintenance program can be initiated without increasing the manpower has been the prime cause of failure in many cases.

You can't squeeze by with the present maintenance work force. If workers already have all the work they can handle, they cannot take on additional tasks without neglecting some of their assignments.

The program will not reduce the maintenance cost—again, at least not until it has been in operation for a while and working smoothly. Maintenance cost consists of labor and material costs, which will increase when you are starting a new program.

The program will definitely reduce the overall operating cost, if adequate manpower is employed for a long enough period to produce results. It results in production of more parts per man-hour, which will result in more profits for the company.

IDENTIFY THE NEED

Maintenance managers must look at their own operations to determine whether the computer is the way to go. They must ask questions such as:

How much more are we spending on maintenance than 5 years ago?

Do we have the information we need to plan maintenance operations?

Can we get this information when we need it?

Is it in usable form?

Do we know what it costs to maintain each piece of equipment?

What are company's plans for operations, for equipment?

And finally, will a computer really help?

If the maintenance information system is basically sound but the information is not easily accessible or more information is needed or it is needed faster, the use of a computerized maintenance information system should be thoroughly investigated.

Think about your maintenance systems and how they affect other areas of operation such as production, accounting, payroll, and customer service. Could these operations be made more productive by improving the speed and efficiency of maintenance? If the answer is yes, you could benefit from converting your maintenance system to the computer.

Even though a maintenance department may feel it can't use a computer because it is small, other factors should be considered before making a final decision:

> How long can the plant tolerate a production line breakdown because a certain part was not available (if maintenance stores are not properly monitored, the company may face more costly breakdowns)?
>
> On the other hand, company capital could be tied up in excessive inventory.

STEP-BY-STEP PROCEDURE

Following is an outline of procedures that can help determine whether a computerized maintenance program will pay off in a particular maintenance operation.

Assign Responsibility

The maintenance manager should assign one person (let us call him "coordinator") from the department the responsibility of looking into computerization of maintenance information and preparing a report.

Establish a Study Team

The coordinator should establish a study team (advisory board) to assist him in the investigation.

The team should be made up of the plant engineer, maintenance manager, industrial engineer, and representatives from the data-processing and accounting departments. At a later date, the coordinator may also need advice from the company's legal department when agreements must be made with the software and hardware vendors.

As the coordinator gets involved with the study, he should consult these people, individually or as a team, whenever a difficult question or problem comes up.

By getting your employees involved in the automation process, you break down their resistance to computers and build enthusiasm for the computer as a tool to make work easier and more productive.

Audit Present Maintenance Practices

The coordinator should begin by thoroughly reviewing all the paperwork systems that are presently required, for example:

- The number of work orders processed monthly
- The amount of backlog on the books (is it increasing or decreasing?)
- The total amount of overtime
- The total amount invested in maintenance stores
- The number of times machine repairs were delayed because of nonavailability of parts
- Any other maintenance-related information

Define Needs and Objectives

List all functions that could be computerized:

- Work orders
- Equipment data and history
- Inventory
- Preventive maintenance scheduling
- Manpower planning, and so on

List department plans:

Handle more work

Plant expansion

New production equipment, and so on

Short-range, as well as long-range, objectives should be listed. A typical short-range objective might be to have inventory control computerized within 6-12 months. At that time you would have the facility for printing a current inventory status, reordering reports, etc. Short-term objectives must complement the long-range objectives. Make a list of your long-range objectives over the next 3-5 years. Typical long-range objectives might be to increase overall maintenance productivity and increase control through more timely information. It is important to be realistic when setting your objectives. Don't try to develop and implement a system that is beyond the state of the art.

Good objectives cannot be defined without performing some preliminary work to determine what objectives should be considered. Some sources of this information are: consultant, short courses, trade publications, meetings, demonstration of actual software packages.

Benefits

List benefits that could be realized (cost savings, service ease, support, speed, etc.). Review the audit as described previously.

Determine portion of each function that can be automated.

Record amount of time spent per period on each function.

Calculate automatable time.

Determine how much the automatable time is worth to obtain the cost savings for each function; if better manual procedures could reduce the overall time, reduce savings accordingly.

BENEFITS OF A MAINTENANCE COMPUTER SYSTEM	Range of Savings, percent	Median, percent
Better Scheduling Planning time per job decreases, resulting in more job plans produced, fewer backlogged jobs, and fewer breakdowns and emergency repairs. Job scheduling is more efficient through availability of reserve stores materials. Only jobs with materials available are scheduled, resulting in fewer schedule changes and reduced time spent by personnel waiting for materials. Job planning is improved by the systems support data, which include planner backlog and status report, quick recall of repetitious job plans, and computerized scheduling and historical job analysis.	5 to 12	8.5
Parts Availability More productive time is assured through parts availability. When parts are not available, craft personnel not only idly await new assignments but also many times spend hours attempting to locate or even fabricate parts.	1 to 3	2
Machine Availability Machine production time increases as the computer contributes to reduced emergency repair through the preventive maintenance program. The ability of the system to automatically schedule preventive maintenance puts this valuable program up front and removes it from neglect.	½ to 2	1.25
Stores Inventory The ability of the system to maintain moment-by-moment inventory levels, automatic ordering, and parts cross referencing results in reduced inventory and fewer stockouts. The automatic cycle counts and updates inventory for efficient parts management.	10 to 20	15

Figure 7.1 Benefits of a maintenance computer system. (From R. D. Mitchell and J. Burgess, *Plant Engineering*, August 7, 1980, p. 83.)

Consider extra savings a computer can provide (better records, faster response to opportunities, less dependence on technical help).

Compute savings from better scheduling, parts availability, machine availability, stores inventory, etc. These savings can be computed from the last 1 or 2 years' history. If you do not have any history, use the table in Figure 7.1.

In the process of evaluating the benefits, you should remember not to look at computers as a tool to reduce employment. Instead, they should be viewed as a tool to reduce the rate of future hiring. For example, acquiring a computer system may permit you to continue to operate with one maintenance clerk as your business expands instead of hiring another one. The computer system will, however, allow your one maintenance clerk to get more work done and to do the job faster and more accurately.

Cost Estimates

Obtain rough cost estimate:

Research catalogs, publications, and technical literature.

Call on software suppliers and computer vendors.

Calculate total costs:

Calculate total software cost. Include:

system analysis and definition

program installation and testing

data entry (plant personnel or outside help)

software acquisition cost

installation cost

cost of modification, if any

training cost

operating cost

software maintenance cost

TABLE I. IMPLEMENTATION COST ESTIMATE

Computer Costs	Cost per Year, dollars First	Second	Third
Systems analysis and definition	8,000	-----	-----
program installation and testing	10,000	-----	-----
Data entry (plant personnel)	15,000	-----	-----
Hardware	5,000	-----	-----
Software cost and maintenance	5,000	-----	-----
TOTALS	43,000		

TABLE II. SAVINGS FROM MAINTENANCE FUNCTIONS

	Savings per Year, dollars First	Second	Third
Labor productivity			
Improvement through better scheduling	51,000	51,000	51,000
Improvements through parts availability	12,000	12,000	12,000
Stores inventory			
Reduction of stock levels	67,500	-----	-----
Reduction of carrying cost	6,750	6,750	6,750
Machine availability			
Increase in machine throughput	34,375	34,375	34,375
TOTALS	171,625	104,125	104,125

TABLE III. NET MAINTENANCE SAVINGS

	Savings per Year, dollars First	Second	Third
Implementation cost	43,000	-----	-----
Operating Cost	12,000	12,000	12,000
Total costs	55,000	12,000	12,000
Gross Savings	171,625	104,125	104,125
Net Savings (Gross savings - costs)	116,625	92,125	92,125

Figure 7.2 Implementation of cost estimate—blank form.

TABLE I. IMPLEMENTATION COST ESTIMATE

Computer Costs	Cost per Year, dollars First	Second	Third
Systems analysis and definition			
program installation and testing			
Data entry (plant personnel)			
Hardware			
Software cost and maintenance			
TOTALS			

TABLE II. SAVINGS FROM MAINTENANCE FUNCTIONS

	Savings per Year, dollars First	Second	Third
Labor productivity			
Improvement through better scheduling			
Improvements through parts availability			
Stores inventory			
Reduction of stock levels			
Reduction of carrying cost			
Machine availability			
Increase in machine throughput			
TOTALS			

TABLE III. NET MAINTENANCE SAVINGS

	Savings per Year, dollars First	Second	Third
Implementation cost			
Operating Cost			
Total costs			
Gross Savings			
Net Savings (Gross savings - costs)			

Figure 7.3 Implementation of cost estimate—example.

Calculate total hardware cost. Include:

hardware

maintenance

supplies

Include every application cost you can think of.

Determine Net Benefits

Subtract total fixed cost from total benefits to find net benefit. Use the form in Figure 7.2 to compile all this information. An example of the analysis is shown in Figure 7.3.

ROI

Perform standard ROI (return on investment) analysis based on operating the system for 3-5 years. If a computer looks attractive, continue; if not, give serious thought to holding off.

Before the ROI analysis is performed, consideration should be given to intangible benefits. Every good economic decision maker gives consideration to the intangible benefits before making the final decision. Intangible benefits are those that cannot be quantified but are important to the decision maker. The computerized maintenance management system has many benefits to which we cannot apply an estimated dollar value accurately. Some of the benefits include: (a) Historical records can be quickly recalled for reference. (b) Historical records accumulated in the data files can be periodically analyzed to help improve performance in the future.

CONCLUSION

This chapter presents an outline of procedures, after the need is identified, that can help determine whether a computerized maintenance management system will pay off in a particular maintenance operation. With constant reduction in hardware costs and growing competition in the maintenance software business, it may not be difficult to justify a system for most maintenance operations. (See Figure 7.4.)

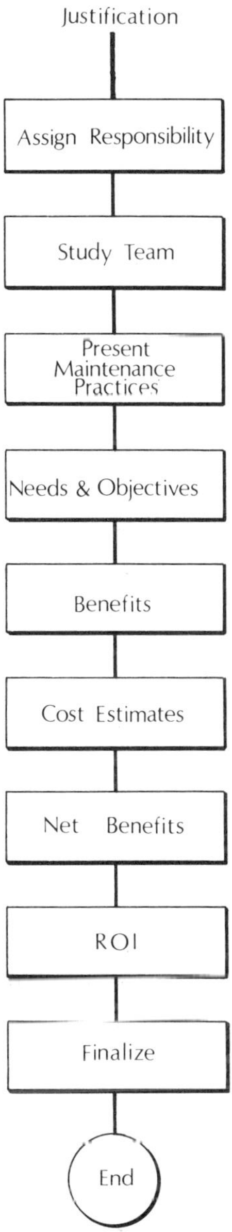

Figure 7.4 Justification—step-by-step procedure.

8

Implementation

INTRODUCTION

Management's responsibilities in implementing the computerized maintenance system involve four activities: motivating employees, planning the installation, supervising the actual installation of hardware and software, and follow-up.

MOTIVATING EMPLOYEES

Employee motivation is an intangible, but important, element in computerizing any application. With your people's enthusiasm and cooperation, the computer can be a productive business tool. Without their support, you won't get the most out of your new system.

Here are some points to consider in getting your employees enthusiastic about automation:

Management Involvement

Your involvement in this new system directly affects your employees. A positive attitude on your part is the most important element of success, because that attitude spreads to your workers. If your initial enthusiasm and commitment fade before you've installed the first application, so will the enthusiasm and cooperation of your employees.

Your people need directions. Let them know what to expect from the system and what is expected of them. Make sure your employees know that you don't expect everything to work perfectly right away. It takes time to adapt to any new system. Let them know that, as with anything new, there will be rough spots. And let them know that they can come to you for help.

Keep up with the applications you install. Determine whether you are getting the benefits you sought. Look for ways in which the system can be improved. Solicit advice and ideas from the people who work on the computer—they will get to know the system better than anyone.

In the end, your attitude—and involvement—determine whether your employees embrace automation or reject it.

Competent Supervision

Appoint someone to supervise the entire operation. This person must be familiar with maintenance, have the support of management, and be a good leader.

Involve Employees

At first, don't expect your employees to share your enthusiasm in automating your company's maintenance functions. Most of us feel uncomfortable and threatened when we face something new. Fear often shows itself as resistance to change, and this resistance can slow down progress.

Prepare your employees for this change and be sensitive to their personal needs and concerns. Get them involved in the automation process as early as possible and keep them fully informed. There are several ways you can encourage good communication:

Ask for employee suggestions. Give employees the opportunity to air their feelings.

Remember that communication is a two-way street. Listen to (and carefully consider) your employees' suggestions, opinions, and complaints.

Periodically call the maintenance staff together to discuss progress that has been made and review any problem areas.

Planning

Planning for the installation of the system is very important. Time spent in planning pays big dividends when you install a computerized system. There are two phases to the installation: installing the computer (hardware) and installing the software. Both the software and hardware vendors provide user's manuals with their products; the manuals contain step-by-step instructions for installing the product.

INSTALLING THE COMPUTER (HARDWARE)

Here is a brief overview of the steps involved in installing a computer:

1. Schedule your computer installation. Set up a schedule for ordering, receiving, and setting up your hardware.

2. Find a place to put the computer. Keep in mind that the computer doesn't need a lot of space, but an adequate work area is helpful to people who use the system. Make sure you put the computer in a place where it is safe and the information stored in the system can be kept confidential.

Here are a few things to think about when selecting a computer location. These factors can greatly affect the comfort and efficiency of the computer operator:

Keep the printer close enough to the keyboard and video display so that the user doesn't have to get up to turn on the printer or attend to paper problems.

A comfortable chair has obvious benefits.

Adequate electrical outlets should be available.

There should be adequate work space to the left and right of the computer for books, manuals, maintenance documents, and so forth.

3. Select the application software. You have already decided by now which maintenance software to acquire. Now is the time to acquire it.

4. Order general supplies. You need supplies such as diskettes, computer paper, and printer ribbons.

5. Receive and set up your computer. When you receive your computer hardware, unpack the manual and read the installation instructions carefully before you set up the system. Just take things one step at a time and you will get it done. If you have problems, the vendor can often answer your questions over the phone.

After your computer is set up, install the operating system on the computer. Again, the user's manual gives instructions for doing this. Take your time and follow the instructions carefully.

6. Become familiar with the computer's features and capabilities. (The amount of training you need depends on your previous experience.) Here are the things you can do to get acquainted with the hardware:

Read the manuals.

Take any training programs that come with the system.

Work through practice problems and exercises provided in the manuals.

Run any training diskettes that accompany the software.

Become comfortable with the general operating features of the system before you move to maintenance application software. Knowing how to run the computer makes it easier to install the software and get it working properly.

INSTALLING THE MAINTENANCE SOFTWARE

1. Schedule the Installation. A planning chart can help you schedule installation activities. It is also a convenient way of keeping everyone informed as to what is going to happen and who is responsible for each activity.

Evaluate your schedule as you go along. Make sure you have enough time to complete everything before you begin the next activity. If not, consider revising the schedule.

2. Prepare Master-File Information. There are certain master files that must be entered into the computer (for example, equipment data, inventory master files, etc.).

The software vendor may provide sample forms you can use to collect this information. However, if you already have this information in a well-organized format, you can enter it directly into the computer without first writing it on the forms.

Also, most master files require numbering of the records put into them (examples: equipment maker, part number, etc.). How you number these may affect the printing of reports and listings. So give some thought to how you assign numbers to master-file records. The software user's manual can usually offer some advice in this area.

3. Train Your Employees. Your employees should complete all of the training the vendor supplies with the software. Here are a few tips for making training more productive:

Give your employees enough time to do the training right.

Provide a quiet place where employees can train free from interruptions.

Follow up on training and evaluate its effectiveness.

Provide additional training and assistance for employees who have completed the basic training but are still not comfortable with the system.

Remember that the success of any computerized operation depends, to a large extent, on the training that operators receive.

4. Get Managers and Support Staff Involved. There are managers and support staff outside of maintenance who will not operate the system but still need to become familiar with the reports and other information it provides.

Hold informal sessions to introduce these managers to the computerized system and what it can do for them.

5. Build the Master Files. In step 2, you prepared forms for various master files. The next step is to create the files on the system and enter the information into the computer. The user's manual should provide instructions for doing this.

After your files are entered into the computer, make a backup diskette of the files. Normally, the files will take more than 1 day to enter, so make a backup copy at the end of each day.

6. Set Up Your Operations. Before you begin using your automated maintenance application, make sure your company is set up to accommodate this new, computerized way of doing things. Any changes in your clerical procedures, control procedures, and work flow must be made at this time. Here are some of the controls and procedures you should establish before converting to the computer:

Set up a filing system: Set up and label files for forms and reports.

Organize work flow: If your present maintenance system is a mess, you must organize it before you computerize.

7. Begin Processing. You made it! Sit back and catch your breath awhile. Observe how things are going, and keep the encouragement and praise levels high. Also, be available to solve any problems that may arise.

PROTECTING YOUR COMPUTER FILES

Information—Your Most Valuable Computer Asset

A computerized system represents a significant investment for most companies. The computer hardware and software must be physically protected in much the same way you protect your inventory, your money, your equipment, and your personnel. But the information stored in your computer may be more valuable to you than the computer itself.

Think about the importance of your maintenance records. What would happen if these records were lost, stolen, burned, or tampered with? How long would it take to reconstruct the information and how much would it cost? Would your business be able to recover quickly from such a disaster? What damage would it do if your competitor was able to gain access to your records?

There are ways that business records stored on computers can be lost or stolen. Let us take a closer look at the problem and how you can deal with it.

Destruction or Loss of Computer Files

Here are some of the ways in which you can lose business information stored on the computer:

- Hardware malfunction
- Programming errors
- Processing errors (controls not maintained, operator mistakes, improper error correction)
- Fire, flood, and other natural disasters
- Improper handling and storage of diskettes

Protecting Your Computer Files

Fortunately, there are a number of steps you can take to protect your computer files:

Scheduling and Rotating Backups

The best way to protect your computer files is to regularly copy them onto diskettes. These copies are called backup diskettes.

If your original files are destroyed, you can use backup diskettes to restore the files and then bring them up to date. This eliminates the need to reenter all your files; you only have to reenter the jobs performed since you made the last backup diskettes.

Here are some general guidelines for making backup diskettes:

Back up your information at the same time every day. This makes backing up a normal part of your work routine.

When you back up your company's records at the end of each month, save the monthly backup diskettes for at least 2 months before using them again.

Prepare two separate sets of backup diskettes. If one set is damaged or destroyed, you still have a good backup file from which you can recover information. Keep the second set at a location away from your office. (If you have a fire, backup diskettes kept at the office can burn with the original files.)

Also make backup copies of your application programs and operating system. Remember, without the application programs, you can't get into your computer files because the computer depends on the application software to process your information. Keep the backup copies of your application and operating system in a safe place.

Some software products come protected and can't be copied. If that is the case with your software, order an additional set of software diskettes from the manufacturer as soon as possible. Then, if the originals become damaged, you won't have to wait several days or weeks for replacement diskettes to arrive. You should check the license agreement for the software that you are considering.

Finally, keep a log that shows what information is on the backup diskettes and which set of backups is most current.

Care, Handling, and Storage of Diskettes

Diskettes come permanently encased in a plastic covering that protects the diskette's magnetic surface. Never touch a diskette where its surface shows through the slots in the covering. Oil in a fingerprint or from hand lotion can damage the diskette and the information it contains.

Each diskette comes in a paper envelope from which it can be removed for insertion into the disk drive. Keep diskettes in their envelopes when they are not in use. Do not store diskettes in the disk drive. You can store diskettes in the box they come in or in special plastic diskette cases (available at computer stores or from your diskette vendor).

Keep diskettes away from magnets such as those used in hi-fi speakers, power transformers, telephones, and electric motors. Magnetic fields can alter or erase information stored on diskettes.

Make sure you do not place diskettes on or near an area where they can come in contact with food or beverages. Anything spilled on a diskette can cause severe damage.

Diskettes are also sensitive to temperature extremes, so be sure to keep your diskettes out of direct sunlight and away from other sources of heat. They may be stored at temperatures ranging from 40 to 125° F. Keeping diskettes in a car or trunk can damage them.

Pressure from pencils and ballpoint pens can damage diskettes. To avoid this problem, use a felt-tip pen to write on diskette labels. In addition, do not attach paper clips to diskettes because the clip can dent or scratch the disk's surface.

Diskette Labeling

Be sure to put a label on each diskette on which you store information or programs. Put the label on the top portion of the plastic covering of the diskette, not on its protective envelope.

Write as much descriptive information on the label as is needed to identify the diskette's contents. Try to keep it brief, however, so that you can read the diskette's contents at a glance. Always write the date and time on the diskette label so you know which copy is current.

Diskette Filing

There is no "right way" of filing diskettes. Simply choose a method you are comfortable with.

Fireproof Vaults

Lock your diskettes in a vault or safe that is fireproof and heatproof. A vault not only protects your files from fire and heat but also prevents theft and keeps your records confidential.

A vault can only guarantee a few hours of protection against fire. If a fire lasts longer than the vault's limits, you run the risk of melting your diskettes. An on-site vault offers limited protection, so be sure to have a second set of backup diskettes stored in an off-site location.

Keeping Computer Files Confidential

The steps you can take to keep computer files confidential are similar to the precautions you have used to keep company secrets in the past. Let us take a look at a few of these security measures:

Computer Location

Keep the computer in a location where it is visible to supervisory employees at all times during working hours. Then you can easily see who is using the computer and what he is doing with it. After working hours, lock the room where the computer is located.

Computer Reports

Store computer reports containing sensitive or confidential information in locked cabinets. Distribute the reports only to those employees who need to see them.

Computer Trash

Incorrect or outdated computer reports are often thrown away without being shredded or even ripped up, but your "computer trash" can reveal confidential company information. Be sure to shred all your confidential reports.

Also, be sure to erase the information contained on old diskettes before you reuse them or throw them away.

Passwords

Your application software should provide passwords to ensure that only authorized employees—those who know the passwords—have access to your computer files. Make sure that passwords are kept secret and changed often. And give the passwords only to those employees who need to have them.

Telecommunications Security

If you use your computer to send information over the telephone, you need to prevent outsiders from tapping into your computer files through this telephone connection. This is a technical area and should be discussed with your vendor or technical advisor.

CONCLUSION

The implementation phase should not be taken lightly because a lot of time and effort are required during implementation. In fact, you will devote more time to the implementation phase than any other phase.

Remember to:

Plan the implementation

Train employees

Provide user's manual and other system documentation

Assign employee responsibilities

Prepare input data

Develop operating procedures

Specify report frequency and number of copies

Define security procedures

Delay final payment until you are sure everything is operating correctly

Although the implementation phase of computerization can be a long, frustrating process, the benefits are well worth the effort. If the benefits were not substantial, no one would be using computerized systems. Most users are pleasantly surprised at the capabilities of their systems after implementation is completed.

9

Conclusion

The effectiveness of a computer system depends on how well it is used, and how well it is used depends, to a large extent, on how well it is understood by the managers and workers who have to operate it. Before introducing a computer system, management and maintenance should consider these two problems: (a) the length of time, without any visible progress, between the system design stage and final implementation; (b) simulating what the computer will do so that people will understand it when it does come into operation.

The information generated by a computer is a powerful management tool when there is credibility in its accuracy, its meaning is properly understood, and management has the courage to act on it. Satisfying these conditions involves organization and training, the first essentials of every successful management program. No matter how simple or complex the system is, it will not work unless its objectives have been identified, and individual jobs have been clearly described to ensure that these functions are carried out. It is not enough simply to draw the charts and write the job descriptions. The people involved must then be made to understand the real meaning and purpose of the job they are to do. Formal training is an important starting point, but real understanding comes only from practice in doing the job.

Following the application of computerized maintenance management measures, a company can expect productivity to improve in almost every operation.

Such an accomplishment is impossible when people think they work for an inept boss in a poorly managed company. The bosses must be transformed into intelligent supervisors who can articulate goals confidently, encourage superior performance, and show each worker appreciation of his skills and contributions.

Review the effectiveness of your system 6–12 months after installation. No system is perfect, and by periodically comparing your results to what you forecast initially, you will be able to identify where improvements are needed.

Detect trends or deficiencies in program.

Check that all vital equipment is covered.

Check that all records are kept in the best possible manner.

Review the economics of the program.

A computerized maintenance system is basically an information-processing tool; it can't make decisions. To derive full benefits from a computerized maintenance system, you must use the information generated to help you find ways to improve your operation.

The importance of maintenance software programs in industry is dramatic and will increase with every passing year. No matter how efficient and cost effective your current planned maintenance system is, if it is not eventually computerized and automated, your plant is in danger of being left behind.

Appendix A

System Specifications—Examples

EXAMPLE 1. FUNCTIONAL SPECIFICATION FOR MAINTENANCE MANAGEMENT SYSTEM (INCLUDING INVENTORY CONTROL) SOFTWARE FOR MICROCOMPUTERS*

1. GENERAL

   1.1 Description

   This functional specification describes the required capabilities and characteristics of a microcomputer program for a maintenance management system. The program is intended to meet the needs of most industrial facilities.

---

*Courtesy of Malcolm Pirnie, Columbus, Ohio, and UNIK Associates, Brookfield, Wisconsin.

1.2 Program Objectives

The MMS program is intended to be a tool for assisting plant maintenance personnel in managing the work effort and associated information, storage, and retrieval for the performance of equipment maintenance services. The program will meet the following functional objectives:

- Be readily usable by maintenance personnel who have received a few hours of training in use of the program.
- Perform several routine data management tasks which otherwise would be handled manually, thereby saving time and effort. These tasks include the storage and retrieval of the following types of information:
  - Equipment nameplate and descriptive specification data
  - Work order information for work in progress as well as completed work orders
  - Historical maintenance cost data, including labor and materials costs
  - Spare parts inventory control
- Generate and print work order forms for both corrective maintenance and preventive maintenance work.
- Provide an easy-to-use means of entering information for initiation of work orders and for updating completed work order records.
- Organize and manage an equipment preventive maintenance (PM) program.
- Be readily adaptable to a variety of plant maintenance organizations providing functions which are as comprehensive as possible, yet allowing the individual user departments to utilize only those features which are appropriate for their organization.

1.3 Implementation

The MMS program shall be written in a high-level, standard programming language and compiled for use on the IBM Personal Computer and other compatible microcomputers.

1.4 Minimum System Requirements

The program shall run on an IBM Personal Computer and other compatible microcomputers equipped with either two floppy disk drives, or one floppy disk drive and one hard disk drive. However, for practical considerations (such as storage required for the complete database or speed of use) it may be appropriate to recommend not using the program without a hard disk drive except for demonstration or relatively small maintenance department applications.

2. PROGRAM FUNCTIONS

2.1 Equipment Data

2.1.1 This module of the program shall have the ability to store and retrieve descriptive information and enter for each individual unit of equipment. The user shall be able to add new equipment records (with appropriate level of user access), and print existing equipment record information.

The user shall have the ability to access an existing equipment record by specifying its unique equipment number, or by searching through sequential records starting at a specified equipment number.

The user shall also have the ability to search for and display or print all equipment records which satisfy various user specified criteria, including: cost center, manufacturer, description, capacity, location, or vendor.

2.1.2 The equipment data records shall include the following fields of information:

Equipment number
Equipment description
Cost center (department or area)
Location
Manufacturer

Model/type
Serial number
Capacity
Electrical data:
  Motor manufacturer
  Model/type
  Serial number
  RPM
  H.P.
  Voltage
  Amps
Elapsed run time
Date installed
Cost
Weight
Manufacturer contact
Manufacturer phone
Vendor contact
Vendor phone

2.1.3 The program should generate the following reports:

Equipment Cost Summary
Shall include equipment number, description, and cost.

Equipment Location
Shall include equipment number, description, and location.

Equipment Listing
All the information on an equipment should be printed.

Elapsed Run Time
All these reports to be printed for a particular equipment or range of equipment. The selection shall be made by equipment number or any other field. Reports will always be sorted by the equipment number.

2.2 Work Order System

2.2.1 This module of the program should have the ability to originate new work orders, print authorized work order forms with instructions and other relevant data for use by the maintenance crafts personnel, accumulate status and cost information for each completed work order, and provide the ability to store and retrieve all associated data for work orders. The user shall be able to inspect new active, overdue, and completed work order records on screen, update and log completed work orders, cancel work orders (with the appropriate level of user access), and print out various summary reports of work orders.

The program shall have the capability to search work order records and display or print all work orders which satisfy various user-specified criteria, including: equipment number, cost center, location, date (specified range), labor cost, material cost, total cost, and current status (e.g. new, active, overdue, complete). The user shall be able to specify multiple search criteria for these reports.

2.2.2 The program shall provide for origination of new maintenance work orders only by users with the appropriate level of access (or authorization). Required data to originate a work order shall be entered using screen display form similar in appearance to the printed work order form. (See Section 2.2.3.)

The equipment name and location shall be automatically displayed or printed based on the equipment number. Error checking will be provided for an equipment number which does not exist.

When a maintenance work order is originated, the user shall have the ability to display or print a list of all PM jobs for the unit of equipment including the most recent performance date and the expected next due date. If desired, the user shall have the ability to originate a PM work order for that equipment unit to be scheduled simultaneously with the maintenance work order. The work order forms which are printed for use by the maintenance crafts personnel will display all appropriate information and instructions. Blanks will be provided to be filled in by maintenance personnel after work completion with the following information: actual labor hours, materials used, description of work performed, completion signature, date, and approval.

2.2.3 The work order form which is displayed and printed shall include the following fields of information:

Work order number
Priority
Work category
Name of requestor
Date of request
Equipment number
Equipment location
Work requested (description)
Estimated and actual labor hours by craft (up to 6 crafts)
Date issued
Date required
Approval (name/signature)
Completion (name/signature)
Completion date
Status code
Planned and actual materials used
Work performed (description)
Other instructions/remarks

2.2.4 Reports

The following reports should be generated:

Work Order Labor Report

Work order number, craft, description, estimated hours, actual hours, labor cost, productivity percentage

Work Order Material Report

Work order number, quantity, part number description, unit cost, extended cost

Work Order Cost Summary

Work order number, labor cost, material cost, total cost

Active Work Orders

Listing of all open work orders

Overdue Work Orders

Listing of all overdue work orders

The majority of these reports could be requested for specific equipment and period. The first 3 reports can be generated for a specific equipment and period to reflect repair or breakdown cost summary. The labor reports can be requested by "craft" also.

2.3 Preventive Maintenance

2.3.1 This module of the program shall have the ability to maintain records for and manage the preventive maintenance (PM) work and inspections for each unit of equipment. The user shall be able to specify frequency for work to be performed. The program shall have the ability to schedule PM work and to automatically generate PM work orders.

The user shall be able to inspect PM work records using the same searching and display/print capabilities as described in Section 2.2.1 for work orders. The user shall be able to log completed

PM work orders, update or revise PM work records, delete PM work records (with the appropriate level of user access), and print out various summary reports of PM work records.

The user shall be able to define and enter PM job tasks for specific units of equipment (by equipment number), and shall be able to assign duplicate PM jobs to identical or similar units of equipment without reentering all necessary information describing the PM job.

2.3.2 The program shall be capable of scheduling PM work orders based on the user-defined frequency for each PM job. The frequency shall be determined by either calendar time or equipment usage, as specified by the users for each equipment unit.

For calendar time scheduling, the program shall compare the elapsed time since the last work performed for the PM job versus the time interval (frequency) specified for the PM job task.

For "run-time" scheduling, the user shall enter equipment elapsed run-times on a periodic basis, using a screen entry format provided by the program. The PM jobs will be scheduled by the program when the elapsed equipment run-time reaches a user-specified percentage of the defined frequency for the PM job. (This percentage will be specified initially in the MMS program configuration.)

The program shall be capable of storing the current total elapsed run-time for each unit of equipment. The program shall provide for entry of run-times by the user either with a separate equipment run-time entry procedure or when logging completed maintenance or PM work orders.

The program shall allow text data entry to provide the ability to combine multiple PM job tasks for the same unit of equipment and schedule all such jobs for the same PM work order.

PM work orders may be originated by any of the following methods:

Automatic scheduling based on frequency, as described above.
Manual origination by the user (with appropriate level of access) from within the PM module of the program.
Manual origination by the user from within the work order module of the program (as described in Section 2.2.2)

2.3.3 The PM work order form which is displayed and printed shall be similar to the maintenance work order form. It shall include the following fields of information:

PM work order number
Work/inspection description
Equipment number
Equipment location
Date issued
Date required
Completion (name/signature)
Completion date
Status code
Estimated and actual labor hours
Other remarks

Provision shall be made to enter the material and/or tools required to perform a PM function.

2.3.4 Reports

Preventive Maintenance Schedule:
Equipment number, job number, craft, frequency, description
This information is printed for all PM functions that are due

PM Master Schedule:
Listing of all PM records in the file

2.4 Equipment History

2.4.1 This module of the program shall have the ability to access previous maintenance history data for all equipment units entered in the program's database. The user shall be able to display and print summary reports showing labor, material, other (outside service contract, etc.), and total costs. The program shall have the ability to present summary costs by user-specified criteria, such as equipment number or cost center, and by specified range of dates.

The program shall also have the ability to present summary detail reports showing materials used and cost, and labor hours and cost by craft. These reports shall be displayed or printed by user-specified criteria and range of dates.

2.5 Inventory Control

2.5.1 This module of the program shall have the capability to store and retrieve appropriate information for spare parts and materials in the facility's inventory. The program shall automatically update quantities on hand after each transaction.

The user shall be able to access the inventory master files in order to add new items, inspect existing inventory records, update or revise existing records, or delete existing inventory records (with the appropriate level of user access). The user shall be able to enter inventory transactions using a separate transaction function.

The program shall provide the capability to display or print various inventory reports, including: master inventory listing, inventory on hand, inventory reorder report (items below reorder point), and inventory activity report (listing of transactions). The program shall have the ability to present inventory reports by user-specified criteria, such as vendor, unit cost, and date of transaction.

2.5.2 The program shall have the capability to automatically adjust inventory-on-hand quantities, using the inventory item numbers and quantities when completed work orders are logged in.

This function can be activated or deactivated for all transactions at the initial setup of the program.

2.5.3 The inventory records which are displayed or printed shall include the following fields of information:

Item number
Item description
Location
Vendor number(s)
Unit cost
Quantity on hand
Reorder point
Reorder quantity
Lead time

2.5.4 The program shall have the capability of charging a part to a specific equipment and/or account number. It should also maintain a history of parts usage.

2.5.5 The program shall have the "order entry" capability and provide the option to automatically generate a purchase order for parts that are below reorder point.

2.5.6 The program shall have the capability to maintain an index file of all vendors. The user shall have the ability to call up a display of the vendor(s) information directly from any master inventory record, based on the vendor number(s) stored for that item. A master listing of all vendors' records shall also be available for display or printing.

Each vendor record shall contain the following fields of information:

Vendor number
Company
Contact
Phone number
Address
Local representative
Local phone
Local address

2.6 Control Program

2.6.1 This module of the program shall have the ability to define certain initial program configurations, control password access by users to the program, define labor crafts and wage rates, and provide the ability to archive old database information.

Use of this module of the program will be restricted to users with the highest level of program access.

2.6.2 The initial configuration routine will provide the capability for maintenance management personnel to initially set up the program, in order to conform with individual organization needs. The following items shall be configurable (at the user's option):

Define equipment numbering system: number of characters, type of characters (alpha, numeric, or both), sequence of character types. In addition to the above, two subfields may be defined within the equipment number: one or

more duplication characters (used to differentiate multiple identical units).
Define cost centers (department or area within the facility) by character code as used in the equipment numbering system and name.
Specify whether or not estimated labor hours are to be printed on work order forms for use by crafts personnel.
Specify whether or not Inventory Control module is automatically linked to Work Order and PM modules (whether or not materials usage automatically updates inventory-on-hand quantities).
Define work order priority codes, work order status codes, and work category codes. Program will error check for illegal code entries for these items. Specify percentage of equipment elapsed run-time frequency to be used for scheduling PM job tasks.

2.6.3 The Control Access routine will provide the ability to assign user passwords and associated levels or user access to various functional modules within the program. The levels of access will include at least the following (from lowest to highest level):
Inspect data (no access to costs)
Add and update data
Inspect data (with access to costs)
Delete data
Originate work orders
Control program
Control access

The program shall provide the ability to perform the following tasks:
Assign new users
Change passwords

Change access level
List users and their password and access
Delete users

2.6.4 The Define Crafts routine shall provide the ability to define craft codes, descriptions, and calculations performed by the program. The user shall also be able to update labor rates as they change. Updating labor rates shall not affect previous labor cost calculations for historical mainance cost data.

2.6.5 The Archive Data routine shall provide the ability to transfer (copy) data from the current disk drive used for database storage to another disk drive. After questioning the user whether original files are to be removed and after entry of a special password, the program will then erase the data files which have been properly archived. The program will also provide the ability to restore archived data files to the program's active database.

3. USER INTERFACE

3.1 Screen Displays and Menus

The display screen shall have a consistent format and appearance at logically similar points throughout the program. The program shall be menu-driven, with the user selecting options from various menus displayed on the screen in order to perform the various program functions.

The menus will be arranged in a hierarchical structure, with the Main Menu providing access to major program functions, and other menus under each main menu selection as necessary to access specific functions and features. Menus shall be used to present several alternate choices to the user, and shall not be used to present a simple question.

3.2 Entry Format
Screen entry forms shall be similar throughout the program. Where possible, text and entry fields will be separated by blank lines to improve readability. The same screen-driven entry format and procedure shall be used throughout the program.

3.3 Keyboard Usage
The number of keystrokes required to perform program functions shall be minimized to the greatest extent possible. The effects of keys shall be consistent throughout the program.

If used, function keys shall have the same meaning wherever they are used in the program. When not used, function keys shall be disabled.

Erroneous keyboard entries, or keystrokes which are not permitted at a specific location within the program, shall cause a beep and result in no action.

4. PROGRAM SECURITY

4.1 User Access and Log-in
The program shall be provided with a complete system of user access controls to provide MMS data security. A log-in procedure shall be provided to start the program. The log-in routine shall require user to enter name and then password. The password shall not be echoed to the screen display as the user enters it.

Control of user access and passwords will be provided as described in Section 2.6.3.

The program shall check the access level required for use of each program module and major function, and display appropriate message if access is denied to the current user.

5. DOCUMENTATION

5.1 User's Manual

Written documentation shall be provided for the program to be used as a reference and User's Manual. The documentation shall be a stand-alone reference. A sufficient level of detail shall be provided and the text shall be written in easy-to-understand language, such that an employee of average intelligence may be able to use and perform the major program functions without outside help. The documentation shall be provided with facsimiles of display screens and examples. Each program function or variation shall be explained.

EXAMPLE 2. GENERAL FUNCTIONAL SPECIFICATIONS*

1. General

1.1 Provide functional system overview.
1.2 Provide flow charts depicting information processing (files, programs, subroutines, etc.) including details on all calculations, formulas, etc.
1.3 Comprehensive and easy-to-understand on-line user operating instructions.
1.4 Adequate "help" documentation available on-line to guide user through system operation.
1.5 Comprehensive and easy-to-understand documentation of how the system works from a data-processing standpoint.
1.6 Reports/screens easy to understand.
1.7 CRT transaction highlights and denotes operational/required fields.
1.8 Menu-and-submenu-driven system.

---

*Courtesy of The Trane Company, La Crosse, Wisconsin.

1.9 System software and application package support available to La Crosse, WI, Scranton, PA, Lexington, KY, Rushville, IN, and Salt Lake City, UT.

1.10 Policy for resolution of software errors—timing and escalation.

1.11 Vendor technical resources to support software installation and debugging.

1.12 Vendor has technical resources to support data-processing training, if requested.

1.13 Vendor has technical resources to support user training, if required.

1.14 Software package fully operational "off the shelf" and has been demonstrated to be essentially bug-free.

1.15 Software packages have been fully and successfully implemented in several manufacturing businesses.

1.16 Vendor software package installed base and internal support and enhancement plans are sufficient to indicate vendor's long-term commitment and staying power.

1.17 Software upgradability policy.

1.18 Right to copy or "execute only" license fees available for satellite plants. (See 1.9.)

1.19 Provide concurrent on-line transactions and batch reporting.

1.20 Software package support "training/test" database.

1.21 All dates on screens displays and reports must be calendar dates (mmddyy or yymmdd with option of displaying shop dates).

1.22 Use names along with numbers where applicable.

1.23 Provide daily audit trail of transactions—type, document number, quantity, before/after value—hard copy report—on-line access to transactions since last report, by item number.

1.24 Detail report-writing capability of system.

2. Technical

2.1 Must be capable of operating/connection to an IBM SNA/SDLC network.

2.2 Application package should be capable of interfacing with a variety of data collection devices.
2.3 Number of terminals—up to 15.
2.4 Systems must be flexible to accommodate organization changes and growth in volume and number of users.
2.5 Can the system screens be customized?
2.6 What types of security do you use to prevent unauthorized on-line access and updates to databases?
2.7 Is there any function of the system that prohibits the availability of the on-line functions? Are any lockout situations possible?
2.8 Which public accounting firm has attested to the correctness and availability of any of your installations?
2.9 Define audit trails, internal controls, and audit reports.
2.10 Can the audit trail be modified independently of the master file?
2.11 What transactions can be secured?
2.12 Can transaction types be delegated specifically to a user or device?
2.13 Does your system interface with any other software other than your own?
2.14 Specify all files and their access method.
2.15 How many batch programs are in the system?
2.16 What benchmark run-time statistics are available?
2.17 How many on-line programs and transactions are in the system?
2.18 What language are the programs written in?
2.19 How is the software supplied?
- Program source?
- Program object?
- Job control language?

2.20 What types of initial database load programs are supplied?
2.21 What is the format of the maintenance updates?
- Object?
- All source?
- Partial source?

2.22 What are the constraints to user modification of the source code?
2.23 Can releases be skipped?
2.24 Are data element descriptions available?
2.25 Was the system originally developed under another operating system?
2.26 Are database updates generally done on-line?
2.27 On what basis do you assess computer utilization and what is that assessment of your system?
2.28 What documentation is delivered with the software?
2.29 Are samples available for review?
2.30 How is documentation updated for new releases?
2.31 What is your policy for resolution of software errors?
2.32 Is phone support provided to research technical questions?
2.33 What vendor technical resources are available to support software installation?
2.34 How many customers do you currently have using this system?
2.35 Does the system have any special hardware requirements?
2.36 Response time of 5 sec or less is required 95% of the time.

16. Plant Maintenance

16.1 The system must be capable of addressing the needs of tools, equipment, parts, and gages (herein referred to as items) used in plant and equipment maintenance and tool and gage control.
16.2 This system will replace the current General Service Requisition (GSR) system as it relates to the Maintenance Department.
16.3 The system must be available to maintenance personnel whenever they will be working (24 hours per day, 7 days per week).
16.4 Data must be accessible by multiple plants in different geographic locations (i.e., La Crosse north/south sides—plant 4/plant 7).
16.5 The system must have appropriate on-line "help" facilities.

16.6 Ability to do generic search on item description.

16.7 Ability to enter work orders on-line, by authorized personnel.

16.8 Ability to view status of open maintenance orders and assign priorities.

16.9 Ability to provide for routine planned maintenance.
- Servicing instructions
- Servicing schedules/work list

16.10 Maintenance personnel should clock in and/or out of each maintenance task, within the maintenance system.

16.11 Ability to interface to Labor and Attendance system for purposes of Payroll and Accounting.

16.12 Must be capable of interfacing to Shop Load System.

16.13 Must be capable of interfacing to "host"-based Purchasing and Receiving systems for placing of orders and receiving orders.

16.14 Requires capability of relating to alternate items.

16.15 All items must contain coding and classification identification.

16.16 Ability to store and maintain (add, change, delete) parent/component relationships (tools, equipment, gages, parts) on-line including continuity checking.

16.17 On-line and hard copy reporting function that includes the following formats:
  Single-level and multiple-level "bill of material"
  Single-level and multiple-level where used
  Search for specific classification of items
  History by item of maintenance performed
  Spare parts/equipment list

16.18 Provide for "obsolete" item coding and retention of obsolete item data for disposition purposes.

16.19 Provide for the ability to relate Trane item numbers to supplier item numbers and access by both numbers, including multiple levels.

16.20 Ability to maintain (add, change, delete) repair operations by equipment including the following:

a. Operation number
b. Operation description
c. Labor type/skill
d. Location where work is performed
e. Labor time
f. Reference to MOP data

16.21 Ability to maintain skills database related to repair operations data.

16.22 Ability to maintain supplier data about what they sell (text) and what we buy. Supplier number should interface to host-based supplier database.

16.23 Planned maintenance should be reflected in the production schedule.

16.24 Breakdowns need to be flagged quickly.

16.25 Maintenance requirements need to be prioritized.

16.26 A history of downtime should be kept on equipment.

16.27 The system should check and monitor response/react time to maintenance problems.

16.28 Cost of maintenance by equipment should be monitored. Break cost down into parts, labor, and other. Need to keep track of what was actually paid for an item in the database.

16.29 The system should check and monitor response/react time to maintenance problems.

16.30 Maintenance problems should be tracked by problem type (electrical, mechanical, etc.) within each machine.

16.31 The system should communicate maintenance requirements directly to the maintenance department as the requirements are known.

16.32 Maintenance requirements should be categorized as to the response expected (immediate, within 4 hr, etc.).

16.33 Based on history, the system should provide an estimate of how long it will take to fix a problem. This information should be available only to authorized personnel.

16.34 When maintenance on a machine is complete, both maintenance and production should confirm that the machine is ready to go back into production.

16.35 Inventory management of items is needed on-line:
  a. Items in stock or not
  b. Sources of items
  c. Lead time and costs of items up-to-date
  d. Order point triggers purchase order notice for manual review
  e. Is rebuild source available?
  f. Cross-reference of parts to machines
  g. Issues and receipts
  h. Multiple storage locations
  i. Interface to Purchasing and Receiving (host)
  j. Pick lists
  k. Cycle counting
  l. List of all items stocked, by location
  m. Allocations

16.36 Ability to keep track of where equipment is moved from and to based on issuance and completion of a work order to do so.

16.37 Miscellaneous management batch reporting, on demand by end user.

16.38 The system must be capable of handling the following volumes;

80,000-100,000 items
150-200 work orders per day
100-150 maintenance personnel
500-1000 transaction per day
60-100 receipts per day
20-30 purchase orders per day
185-225 open work order backlog
3-day average life of a work order

Appendix B

Software Purchase Agreement

AGREEMENT effective ________________________, 1985 by and between XYZ (Buyer), and UNIK Associates, with its principal place of business at 12545 W. Burleigh, Brookfield, Wisconsin, 53005 ("UNIK").

The parties agree as follows:

1. UNIK shall sell the software product "Advanced Maintenance Management System (MMS II)" to BUYER.

UNIK is developer and publisher of this software product MMS II. MMS II is copyrighted under the name of UNIK Associates.

2. *Payment*: UNIK shall be paid a fee of Five Thousand, Nine Hundred, Ninety-five Dollars ($5995.00) for the program,

documentation and training. Payment by BUYER to UNIK shall be as follows:

a. Initial payment of Three Thousand Dollars ($3000.00) with purchase order.

b. Final payment of Two Thousand, Nine Hundred, Ninety-five Dollars ($2995.00) upon delivery of program and completion of training.

3. BUYER agrees that this software product is to be used only by BUYER at ______________________________.

4. UNIK shall provide, if necessary, technical assistance and support to buyer for the MMS II product over the telephone after completion of training.

5. Any programming errors found by BUYER within a period of twelve (12) months after completion of the product shall be fixed by UNIK at no charge.

6. *Modification*: If requested by BUYER, UNIK shall make any reasonable modifications to the MMS II product. The fees and completion date shall be agreed upon by the parties prior to the initiation of the modification.

7. A copy of the source code will be retained in the office of ABC ("Escrow Agent"). In the event UNIK shall become bankrupt or insolvent, the source code held by the Escrow Agent shall be transferred to buyer.

8. *Patents and Copyrights*. UNIK shall, at its own cost, defend BUYER from any claim, suit, or proceeeding brought against BUYER based on a claim that the use or transfer of any products delivered hereunder constitutes the infringement of any patent or copyright in the United States. UNIK shall pay any costs and damages awarded against BUYER in such action which is attributed to such claim, provided that BUYER notify UNIK promptly in writing of any action, claim, suit, or proceeding brought against BUYER and is given full authority and information and assistance for the defense thereof. UNIK shall hold BUY-

ER harmless from any such claim, suit, or proceeding brought against BUYER.

IN WITNESS THEREOF, the parties have executed this Agreement under seal as of the day and year first above written.

By______________________________
UNIK Associates

By______________________________
BUYER

Appendix C

Vendor Questionnaire

Vendor Name ______________________________
Address ______________________________
Telephone ______________________________
Number of Years in Business ______________________________
Number of Systems Installed ______________________________
Number of Systems Up and Operating ______________________________
Year First System Was Installed ______________________________
Location of Current System Users:

Contact Person	Company Name	Address	Telephone
________	________	________	________
________	________	________	________
________	________	________	________
________	________	________	________

Bibliography

Allerton, L. J. (1983). Controlling maintenance costs via a computer. *Cost & Management*, November–December, pp. 45–49.

Allman, G. W., and Bottom, J. H. (1980). Scheduling maintenance projects with ease—using a microcomputer. *Plant Engineering*, October 16, pp. 128-131.

Baker, J. T. (1980). Automotive preventive maintenance programs for service industries and public institutions. *Industrial Engineering*, February, pp. 18–21.

Baldwin, R. (1983). Computers—Extending the boundaries of plant engineering. *Plant Engineering*, February 3, p. 48.

Bleuel, W. H. (1981). Maintenance is not an expense—It is a new source of profit. *Plant Engineering*, December 10, pp. 75–78.

Burt, E. (1984). No predictive maintenance. *Plant Engineering*, October 25.

Candy, W. L. (1982). Maintenance groups make and implement own suggestions for improving productivity. *Industrial Engineering*, February, pp. 44-48.

Colebrook, W. (1973). *Low Cost Maintenance Control.* Amacom.

Deriso, J. L., and Lane, J. C. (1983). *Maintenance 7. Operations Cost Study.* AS&U, March, pp. 39-54.

Doll, W. J., and Reid, E. J. (1984). Choosing the right computerized PM system. *Plant Engineering*, August 23, pp. 54-57.

Engineering Staff, Utah Copper Div., Smelter Plant (1979). Instant maintenance data. *Plant Engineering*, June 28, pp. 109-113.

Fitzgerald, J. (1978). How PM relates to overall conveyor costs. *Plant Engineering*, June 23, pp. 65-67.

Galto, E. J. (1982). Computerized maintenance data. *Plant Engineering*, September 30, pp. 57-60.

Harding, G. G. (1973). Mechanized maintenance scheduling. *Management Accounting*, March, pp. 36-42.

Harvey, R. E. (1983). Hi tech production will require hi tech maintenance. *Iron Age*, December 14, pp. 44-50.

Hayo, G. (1985). Managing materials with the use of a computer, 1985 National Plant Engineering & Maintenance Show and Conference.

Hays, G. (1979). Controlling spare parts inventory. *Plant Engineering.*

Hershey, T. M., and Claire, F. V. (1979). Using computer data in maintenance management. *Plant Engineering*, March 8.

IBM Guide to Choosing Business Software, An (1984). IBM Corporation and Banbury Books, Wayne, PA.

Jamison, R. S. (1984). Selecting a maintenance software application package, *Plant Engineering*, March 26-29.

Johnstone, J. (1984). Getting the most from maintenance computers. *Plant Engineering*, November 15, pp. 83-85.

Jordan, H. H. (1984). Selecting software packages for production and inventory management: Caveat Emptor. *Material Handling Engineering*, August, pp. 63-65.

Katzel, J. (1981). Programming for progress—Move toward plant engineering management by computer. *Plant Engineering*, February 19, pp. 84-91.

Katzel, J. (1982). Plant engineering in a high-technology environment. *Plant Engineering*, October 28, pp. 70-75.

Katzel, J. (1983). Telecommunications. *Plant Engineering*, February 3, p. 57.

Lawson, D. L. (1978). Maintenance information reports. *Plant Engineering*, June 22, pp. 121-124.

Lazenby, W. W. (1980). Organizing a maintenance department for optimum effectiveness. *Plant Engineering*, May 1, pp. 83-85.

Lazenby, W. (1981). Setting up a preventive maintenance system. *Plant Engineering*, May 28, pp. 223-226.

Long, G. (1979). Numbering system simplifies maintenance spare parts inventory. *Plant Engineering*, March 22, pp. 137-140.

Mann, L. Jr., and Coates, E. R., Jr. (1980). Evaluating a computer for maintenance management. *Industrial Engineering*, February, pp. 28-32.

Marinello, R. L. (1981). Maintenance costs. *Plant Engineering*, August 20, pp. 58-63.

Marinello, R. L. (1982). Computerizing maintenance information. *Plant Engineering*, February 18.

Martin, G. N. (1979). EDP systems maintenance. *Journal of Systems Management*, September, pp. 18-21.

Meinhold, T. (1983). Electronic controls. *Plant Engineering*, February 3, p. 54.

Miller, F. W. (1979). Gaining a new respectability. *Infosystems*, April, pp. 64-69.

Mitchell, R. D., and Burgess, J. (1980). Justifying a computer for maintenance management. *Plant Engineering*, August 7, p. 83.

Montgomery, R. A. (1985). The true advantage of computerizing preventive maintenance, National Plant Engineering and Maintenance Show and Conference.

Newbrough, E. T. (1980). Reshaping unplanned hours improves maintenance productivity. *Plant Engineering*, March 6, p. 291.

Newbrough, E. T. (1981). Managing maintenance materials. *Plant Engineering*, April 2, pp. 51-53.

Newbrough, E. T. (1982). The three levels of maintenance management control. *Plant Engineering*, October 14, pp. 67-68.

Nolden, C. (1981). Engineering and maintenance services—A $24 million business at the Stanley Works. *Plant Engineering*, August 6, pp. 84-90.

Nolden, C. (1983). People plus computers equal quality performance in plant engineering. *Plant Engineering*, October 27, pp. 50-55.

Nolden, C. (1983). The work order system: Key to effective maintenance management. *Plant Engineering*, October 13, p. 54.

Oliver, W. K. (1985). Evaluating CAE software. *Hydrocarbon Processing*, February, pp. 51-54.

Peele, T. T., and Chapman, R. L. (1984). Determining maintenance manpower requirements. *Plant Engineering*, July 12, pp. 74-77.

Powers, P. W. (1980). Improving maintenance productivity. *Plant Engineering*, September 4, pp. 89-91.

Rastani, J. and Rijo, C. (1985). Software selection made easy. *Infosystems*, March, pp. 92-93.

Revelle, J. B. (1978). Determining when to replace parts at least cost. *Plant Engineering*, November 23, pp. 109-110.

Rissel, M. C. (1984). A highway maintenance management system. *Public Works*, January, pp. 48-50.

Rowan, M. (1982). Why maintenance training is a very good investment. *Modern Materials Handling*, March 5, p. 51.

Ruszkiewicz, J. W. (1980). Maintenance goals and management control. *Industrial Engineering*, February, pp. 22-23.

Shepard, W. F. (1979). Designing a small-plant preventive maintenance program. *Plant Engineering*, February 8, pp. 119-122.

Spector, L. (1983). The impact of high technology. *Plant Engineering*, February 3, pp. 47-59.

Steven, R. R. (1984). Preventive maintenance depends on planning. *Offshore*, April, pp. 50-51.

Teresko, J. (1983). Maintenance erodes the bottom line. *Industry Week*, May 30, p. 56.

Tombarge, C. H. (1984). Controlling maintenance costs. *Plant Services*, November, pp. 65-66.

Tomlingson, P. D. (1979). Types of maintenance organizations—How they developed and why. *Plant Engineering*, April 19, pp. 315-317.

Tomlingson, P. D. (1981). The maintenance planner. *Plant Engineering*, November 12, pp. 123-125.

Tomlingson, P. D. (1984). There's no quick fix with package programs. *Coal Age*, May, pp. 79-81.

Weiss, W. H. (1982). Controlling maintenance materials. *Plant Engineering*, July 8, p. 59.

Weiss, W. H. (1982). High-use maintenance stores areas. *Plant Engineering*, March 18, pp. 252-254.

Westerkamp, T. A. (1982). Preventive maintenance pays for itself. *Foundry M&T*, March, pp. 24-28.

Witt, C. Maintenance software programs gain strength, versatility and acceptance. *Material Handling Engineering*, pp. 180-186.

Wolf, G. (1983). Robotics living with the iron collar worker. *Plant Engineering*, February 3, p. 51.

Wooldridge, S. (1973). *Software Selection.* Auerbach Publishers, Inc., Pennsauken, NJ.

Index